CODE
OF
ESTIMATING
PRACTICE

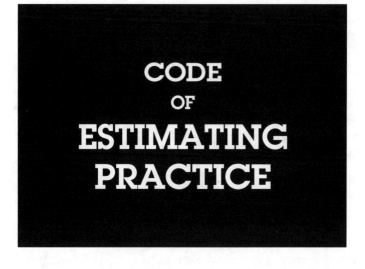

CODE
OF
ESTIMATING PRACTICE

Sixth edition

Endorsed by the Procurement Committee of
The Chartered Institute of Building

The CHARTERED
INSTITUTE OF
BUILDING

Addison Wesley Longman Limited
Edinburgh Gate, Harlow
Essex CM20 2JE, England
and Associated Companies throughout the world

Co-published with The Chartered Institute of Building through
Englemere Services Limited
Englemere, Kings Ride, Ascot
Berkshire SL5 8BJ, England

Fifth edition 1983
Sixth edition 1997

British Library Cataloguing in Publication Data
A catalogue entry for this title is available from the British Library

ISBN 0-582-30279-X

Set by 32 in 10/13 Lubalin Graph Book
Produced through Longman Malaysia, PJB

CONTENTS

CONTENTS

PREFACE TO THE SIXTH EDITION

The first edition of the *Code of Estimating Practice* was produced by the Institute's Estimating Practice Committee in April 1966. It has been revised on four occasions since that time, the last edition being published in October 1983.

This edition of the Code has been totally rewritten and updated to incorporate many important new features which are currently regarded as integral parts of good estimating practice, for main and specialist trade contractors.

When the early editions of the Code were written, the majority of building work was let under what are now called 'traditional' arrangements. In recognition of the greater use of alternative procurement systems, the CIOB has published four Supplements to the *Code of Estimating Practice* as follows:

Supplement No. 1 *Refurbishment and Modernisation* (1987)
Supplement No. 2 *Design and Build* (1988)
Supplement No. 3 *Management Contracting* (1989)
Supplement No. 4 *Post-tender Use of Estimating Information* (1993)

These four Supplements each form a valuable extension to this Code.

In addition, the Institute has recently produced a complementary guide: *Computer-aided Estimating* by William Sher. This gives advice on the implementation of computer systems in an estimating office, and deals with issues such as training, coding systems, libraries and reports.

The increasing use of computer systems in construction is responsible for many of the changes in estimating practice since the last edition of the Code. The way in which estimators use computers affects all subsequent stages in the financial management of a construction project. It is important to be aware of the way data are now handled so that staff can examine the effects of changes to resources and margins. Also many reports which were hand-written are now generated by computer systems in a form suitable for presentation to a review panel. In most cases, however, the basic principles of estimating remain unchanged.

Estimators have always relied on specialist sub-contractors for prices at tender stage, whether the service is for labour or labour and materials. Now that traditional nomination procedures have greatly reduced, it has been necessary to reflect this, together with other trends such as shorter tender periods and more productive working methods. Clearly the comprehensive procedures described in this

Code could not be applied to every tender and where poor documentation and shorter tender periods are present contractors must decide whether to attempt to produce a competitive bid.

For smaller building organisations and trade specialists it would be appropriate to adopt selected parts of this Code and produce more concise forms with the information needed for the business. The *Code of Estimating Practice* provides comprehensive checklists and forms in order to show all the data which can be required during the preparation of a tender.

Estimators' calculations for item build-ups and resources may have been simplified, but the estimates need to be analysed in more detail and to greater accuracy than before if management is to succeed in increasingly competitive markets needing ever smaller margins. For successful bids, the financial success of a project is monitored and measured against the budget produced by the estimator.

Some terms have been changed in this edition to reflect developments since 1983. In particular, the word 'adjudication' has been replaced with 'final review' for the meeting with management at which an estimate is converted into a tender. This recognises the fact that 'adjudication' is a term commonly used to describe the checking of tenders by the client's representative.

Martin Brook, BEng(Tech), FCIOB
Chairman of the Procurement Committee
The Chartered Institute of Building
February 1997

ACKNOWLEDGEMENTS

This new edition was compiled and written by Martin Brook, who was supported by a Steering Group composed of representatives of the CIOB's Procurement Committee. The composition of the Steering Group was:

John Pead	FCIOB, FRICS	*Chairman*
Martin Brook	BEng(Tech), FCIOB	
Doug Elliott	BA, FCIOB, FRICS	
Bill Rabbetts	MCIOB	
William Sher	MCIOB	

The Steering Group wishes to thank Tony Appleton, John Gandy and Peter Harlow for checking the manuscript.

LIST OF FIGURES AND FORMS

TERMINOLOGY

The following are the meanings of the principal specialist terms used in this Code.

Adjudication The assessment of competitive tenders by the client's representative. *See also* 'Final review' below.

All-in labour rate A compound rate which includes payments to operatives and associated costs which arise directly from the employment of labour.

All-in material rate A rate which includes the cost of material delivered to site, conversion, waste, unloading, handling, storage and preparing for use.

All-in mechanical plant rate A compounded rate which includes the costs originating from the ownership or hire of plant together with operating costs.

Approved contractors Approved contractors are those who have demonstrated that they have the expertise, resources, ability and desire to tender for a proposed project. Selection of such contractors is normally by preselection procedures.

Attendance The labour, plant, materials and/or other facilities provided by main contractors for the benefit of sub-contractors, for which sub-contractors normally bear no cost.

Buildability The extent to which the design of a building facilitates ease of construction subject to the overall requirements of a completed building.

CDM Regulations *Construction (Design and Management) Regulations 1994* concern the management of health and safety. These regulations impose duties on clients, planning supervisors, designers and contractors.

Consultants The client's or contractor's advisers on design, cost and other matters. Such advisers may include project managers, architects, engineers, quantity surveyors, accountants, bankers or other persons having expert knowledge of specific areas.

Contingencies A form of undefined provisional sum which is for work which cannot be identified at tender stage.

Cost The estimated cost of the physical production of work. (*Note:* Estimated cost should not be confused with historical cost; historical cost is the cost of construction which is revealed only after the work has been executed.)

Estimate Net estimated cost of carrying out the works for submission to management at the final review meeting.

Cost records Records of historical costs and notes of the conditions prevailing when such costs were incurred.

Domestic sub-contractors Sub-contractors selected and employed by a contractor.

Down time (or standing time) The period of time that the plant is not operating. This may be due to breakdown, servicing time or an inability to operate due to other factors.

Effective rate The rate calculated by dividing a gang cost by the number of productive operatives in the gang.

Employer The building owner or employing organisation, also called the 'client'. The term employer is commonly used in UK construction contracts.

Estimating The technical process of predicting the costs of construction.

Estimator A person performing the estimating function in a construction organisation. Such a person may be a specialist or he may carry out the estimating function in conjunction with other functions, such as quantity surveying, buying, planning or general management.

Final review The action taken by management to convert an estimate into a tender. Also commonly known as 'appraisal', 'settlement' or 'adjudication'.

Fixed price contract A contract where the price is agreed and fixed before construction starts. *The term 'firm price' is used to denote more precisely a contract which will not be subject to fluctuations.*

Fluctuations The increase or decrease in cost of labour, plant, materials and/or overhead costs which may occur during a contract.

General plant Part of a contractor's project overhead calculation for plant excluded from unit rate calculations and which is available as a general facility on site. Durations for general plant are usually taken from the tender programme.

Gang cost A grouping of labour costs to include principal and supporting labour associated with a particular trade. It may also include items of plant.

Head office overheads The cost of administering a company and providing off-site services. The apportionment of head office overheads, to individual projects or as a percentage of company turnover, is decided by management as part of management policy.

JCT *The Joint Contracts Tribunal*, responsible for producing the standard forms of building contract.

Labour-only sub-contractors Sub-contractors whose services are limited to the provision of labour.

Lump sum contract A fixed price contract where contractors undertake to be responsible for executing the complete contract work for a stated total sum of money.

Management Those responsible for the function of general management and having the responsibility for making the decision to tender and for reviewing tenders.

Margin The sum which is required by an organisation, from a project, as a contribution towards its head office overheads and profit.

Mark-up The sum added to a cost estimate, following the final review meeting, to arrive at a tender sum. Mark-up will include margin, allowances for exceptional risks, and adjustments for commercial matters such as financing charges, cash flow, opportunities (scope) and competition. There may be a requirement for main contractor's discount when tendering as a sub-contractor, and Value Added Tax when required in the tender instructions.

Method statement A statement of the construction methods and resources to be employed in executing construction work. This statement is normally closely linked to a tender programme.

NJCC The National Joint Consultative Committee for Building responsible for producing Codes and Procedures for good practice in tendering for construction work. *(This organisation ceased operating in 1996.)*

Nominated sub-contractor/supplier A sub-contractor/supplier whose final selection and approval is made by the client or client's advisers (*see* Prime cost).

Open competition An impartial method of procurement whereby contractors are invited through advertisements to apply for tender documents. The number of tenderers is not usually limited and reputation and ability to execute the work satisfactorily are not always considered.

Pre-qualification The provision by a contractor of information as part of a preselection process. An application by a contractor to be included on a select list of tenderers.

Preselection The establishment of a list of contractors with suitable experience, resources, ability and desire to execute a project, bearing in mind the character, size, location and timing of a project.

Prime cost (PC) When used in bills of quantities and specifications, prime cost means that part of an estimate for work or supply of materials to be provided by nominated sub-contractors or suppliers. Prime cost sums are determined by the client's advisers and detailed in the tender documents. The contractor may also be invited to execute work covered by a PC sum in certain instances.

Project overheads The site cost of administering a project and providing general plant, site staff, facilities and site-based services and other items not included in all-in rates. Also commonly known as: preliminaries, general cost items or general expenses.

Provisional sums *The Standard Method of Measurement for Building Works* (SMM) provides for sums which may be included in tender documents for work which cannot be measured at tender stage. There are two types of provisional sums: defined and undefined.

For *Defined provisional sums*, contractors must be given full information about the nature and extent of the work to enable them to make due allowance for the associated programming, planning and pricing project overheads.

Undefined provisional sums are for work where the contractor is entitled to a valuation which includes any reasonable allowance for programming, planning and preliminaries. Expenditure of an undefined provisional sum may be an event which may lead to a claim for an extension of time under the JCT standard forms of contract.

Risk Additional technical, contractual, financial and managerial responsibilities which form part of the contractor's formal obligations.

Scope Opportunities to improve the financial, commercial or business aims of a construction organisation.

Selective tendering   A method of selecting tenderers and obtaining tenders whereby a limited number of contractors are invited to tender. The tender list is made up of contractors who are considered suitable and able to carry out the work. This suitability is usually determined by preselection procedures.

Standing plant   Plant retained on site which is not working but for which a contractor is still liable.

Temporary works   Resources needed for non-permanent work. Some temporary works such as formwork are measured in a bill of quantity, others such as hoardings are normally excluded from unit rate calculations because they are common to a number of activities and their durations are taken from the tender programme.

Tender   A sum of money, time and other conditions required by a tenderer to complete the specified construction work. For design and build the term tender includes design (Contractor's Proposals) and price (Contract Sum Analysis).

Tendering   A separate and subsequent commercial function based upon the estimate.

Tender documents   Documents provided for the information of tenderers, in order to establish a common basis for their offers.

Tender programme   An initial version of a master construction programme prepared during the tendering period, to enable main contractors to understand the important time and resource considerations of a project in terms suitable to the preparation and submission of a tender. It may be submitted with a tender.

Tender timetable   A timetable for the preparation of an estimate, all necessary supporting actions and for the subsequent conversion of the estimate into a tender.

Working Rule Agreement (WRA)   National working rules for the building industry produced by the National Joint Council for the Building Industry.

1. INTRODUCTION AND FLOWCHARTS FOR ESTIMATING AND TENDERING

This Code of Practice is the authoritative guide to essential principles and good practice in estimating for building work from preselection to the acceptance of a successful tender (see Figure 1.1). It is of value to those members of the building team who are involved with the production and use of an estimate, and to construction students within all disciplines.

The Code distinguishes between *Estimating* and *Tendering* and considers the formal steps to be taken in converting the net cost estimate into a tender at the adjudication stage.

Estimating is defined as the technical process of predicting costs of construction. *Tendering* is a separate and subsequent commercial and managerial function based upon the net cost estimate (see Figure 1.2).

In addition to setting out the basic principles of estimating, consideration is also given to the integration of estimating and management and links with programming, purchasing, commercial management and the construction process.

Management is an important element in the production of an estimate and the subsequent tender. Estimators must have management responsibility within the department responsible for estimating and for managing the production of estimates.

This management role must not be underestimated. Estimators have to ensure that other departments and staff work to their requirements, produce information on time and in the format required, so that effective operating procedures and lines of communication are established between departments to allow the efficient production of estimates.

The initiation and control of enquiries, quotations and programme are an essential part of the estimating process. Estimators must co-ordinate temporary works design and liaise with construction management on methods and available resources. In many organisations estimators also have responsibility for tendering and the subsequent management of projects in the construction phase of a successful tender.

An estimate must be prepared in a way that is explicit, consistent and takes account of methods of construction and circumstances which may affect the execution of the work on a project. A reliable estimate can only be achieved when each operation or item is analysed into its simplest elements and the cost calculated methodically on the basis of factual information.

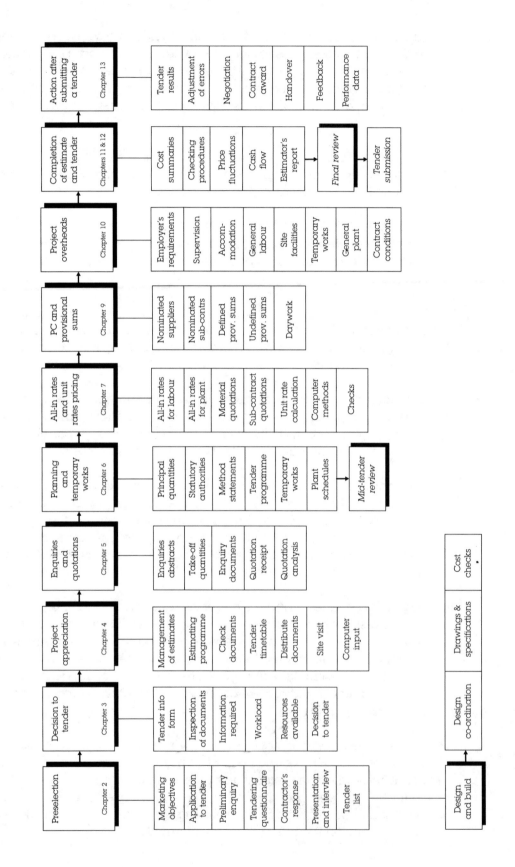

Figure 1.1 Flowchart for estimating and tendering

Figure 1.2 Constituents of a tender

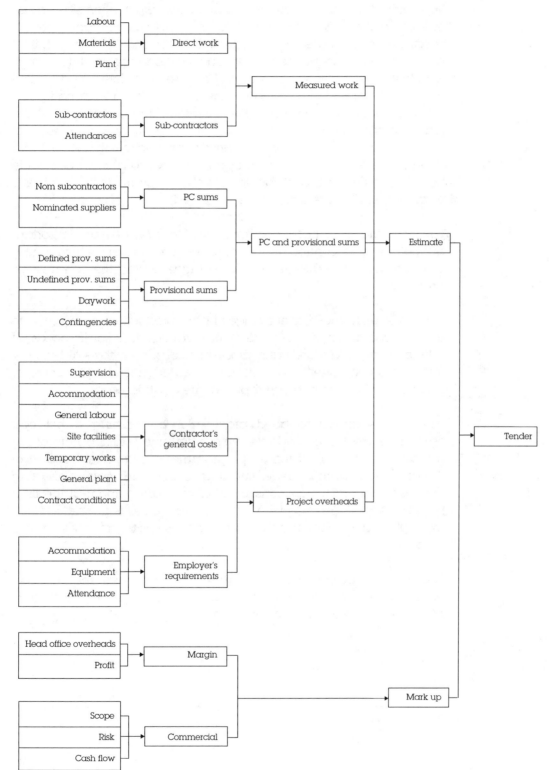

The basic principles of estimating described in this Code and its Supplements will be suitable for wide application by those estimating for various types of building and sub-contract work. In specific circumstances the procedures may be modified or simplified but the essential sequence of tasks and method in which information is obtained is unlikely to change. The use of ordered and logical methods of estimating provides management with information on future demands on resources. The estimate, in conjunction with the tender programme and method statement, provides a means of measuring the financial consequences of any delays that may occur during the execution of a project. Comparisons can be made between estimated and actual levels of productivity, allowing management to take action when required.

The estimate is a base document which provides an important budget for cost control during construction and on which many assessments and judgements will be made during the construction phase.

This Code examines various stages in the production of an estimate and its conversion to a tender. The various functional needs of estimators are considered and procedures described to establish how the basic components of estimates are identified and costs of resources to be employed on a project are established.

Attention is drawn to good practice of establishing a net cost estimate. Project overheads are then determined, taking note of the method statement and tender programme. An important aspect of the estimator's work is to identify areas of risk and matters affecting the costs of a project, for consideration by management at the final review stage. This Code illustrates such matters with priced examples, and demonstrates how the estimate and tender are produced.

The text assumes that bills of quantity have been measured in accordance with the current edition of the *Standard Method of Measurement of Building Works* (SMM).

2. PRESELECTION

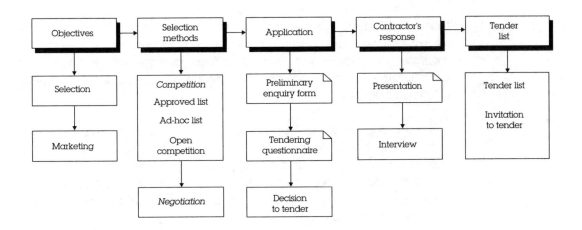

2.1 OBJECTIVES

Preselection is concerned with the establishment of a list of contractors or specialist trade contractors with the necessary skills, experience, resources, previous tender performance and desire to carry out the works, bearing in mind the character, size, location and timing of the project. The contractor will be concerned with planning the workload in his estimating department, satisfying himself that projects suit his particular skills, meet business objectives and ensure that adequate management and resources are available to support the project at the time stated.

Preselection will occur some time before any formal invitations to tender are issued. It is important to have a realistic programme covering the whole of the preselection and tendering period and that adequate time is given for each stage. On major works preselection may occur three months before tender documents are issued to help the contractor plan future work in the estimating department. In the case of smaller contracts a period of four to six weeks is more usual.

The contractor must be given adequate time to evaluate the information provided and prepare the information and data required for preselection purposes. Comprehensive and detailed information will be required and on major works such submissions could take several weeks to prepare.

2.2 COMPETITION AND NEGOTIATION

Contractor selection may be made by competition or negotiation and often by a combination of both. Preselection procedures are usually used to draw up a tender list in order to invite competitive tenders. Where the lowest tender exceeds the employer's budget, the lowest tenderer will usually be asked to assist with a price reduction exercise to enable the project to go ahead. In practice, two contractors are asked to consider the financial savings in order to retain competition.

A contractor may be approached to negotiate a contract price without the need to introduce competitive tenders. In this way a contractor may be selected on the basis of past performance, recommendation, familiarity with the work or most commonly because of a close working or business relationship with the employer or consultants.

Negotiation allows early contractor selection where the design would benefit from the constructor's input. This can reduce the overall project programme, increase buildability and tailor the costs to the employer's budget. The counter argument is that the initial price may be higher and difficult to compare with competitive market rates. There are also procedures adopted by many (usually public) organisations to ensure that goods and services are procured by competitive means particularly where finances are publicly accountable.

2.3   MARKETING AND THE APPLICATION TO TENDER

In recent years contractors have improved their approach to marketing. Invitations to tender seldom arrive in the post 'out of the blue'. At the very least a marketing plan is produced to meet the needs of the business and leads are pursued in an attempt to be included in the initial tender list. It is also necessary to apply for inclusion on approved lists which are maintained by employers regularly using the services of the construction industry.

For a company to succeed it must actively match its resources to the needs of clients in projects where it can work at a profit. A marketing plan will be proactive, bringing construction skills, both technical and commercial, to the early stages of a project's development. It would be misleading therefore to suggest that making a written application to be included on a tender list is sufficient. The contractor must develop close working relationships with his clients and target the core business he wishes to undertake. This is the best way to ensure that there is an even workflow in the estimating department, that tendering costs are kept under control and that a suitable turnover is maintained.

2.4   SELECTION METHODS

Open competition

Open competition is a method which excludes preselection thus allowing any applicants to join a tender list which can be, and often is, beyond a sensible level. This arrangement can no longer be defended for a construction industry concerned about its tendering costs.

Approved lists

This form of selective tendering enables clients to choose tenderers for a project from a standing list of approved contractors, who have been vetted and pre-qualified for various categories of work at an earlier stage. Contractors are asked to apply for categories which are defined by contract value and nature of work.

It is important that clients and clients' representatives monitor and regularly update their lists of approved contractors to:

- Exclude companies whose performance has been unsatisfactory.

- Introduce suitable new companies which can demonstrate the required qualities and abilities.

- Compile the lists in a form appropriate to the class of project.

- Include firms with the financial capacity and stability to carry out the work.

One-off project (ad-hoc) lists

Clients or their consultants frequently create an initial list of suitable contractors solely for a particular project. There will often be a pool of contractors from which to choose, which can be assembled in three ways:

- The list may include contractors who write with an early expression of interest in a scheme.

- An advertisement may be used to invite applications.

- Clients or consultants may draw on their experience or interrogate their database.

Contractors included on an initial list are usually asked to provide information about their financial and technical performance, particularly in relation to the type of work under consideration. More elaborate pre-qualification practices include completion of questionnaires and making presentations to the clients and their consultants. An assessment of a contractor's competence in health and safety is now a statutory requirement prior to the award of a contract and so may be dealt with at the preselection stage.

2.5 PRELIMINARY ENQUIRY

To make a realistic judgement on whether to tender for a project, a contractor must be given the following preliminary information.

Client

- Details of the client, or if a subsidiary company, details of the holding company.

- Full particulars of consultants to be used on the project, including their duties and responsibilities.

- Particulars of proposed site supervision to be provided by the client or consultants.

- A description of the tender documents, their expected date of issue, the period available for tendering, the acceptance period for the tender and time when unsuccessful tenderers will be notified.

- Whether the project, either in its present or a different form, has been the subject of a previous invitation to tender.

- The latest date for receipt of acceptance of invitation to tender.

- The number of tenders to be invited.

Project details

- The location of the proposed works, including preliminary drawings and a site plan.

- Description of the project.

- Approximate cost of the project.

- The date for commencement of the contract.

- The period for completion.

- Details of any phasing.

- An outline of the form of construction.

- Access problems.

- Special operational space requirements.

- Ground conditions.

- Sufficient dimensions and specification details to permit evaluation of the project.

- Details of work to be carried out by nominated sub-contractors, approximate value, and names if known.

- An indication of health and safety issues.

Form of contract and contract particulars

- Form of contract to be used.

- Provisions for liquidated damages.

- Proposed amendments to standard forms of contract and appendix.

- Details of interim payments.

- Retention conditions.

- Bonding requirements.

- Details of insurance.

- Provisions for fluctuations in cost.

- Warranties and insurance policies required.

The contractor must be given prompt notification of any changes in the information provided, both during the period that the invitation is open and subsequently before the despatch of tender documents. It is important that consultants keep to the dates stated in pre-qualification documents regarding the issuing of tender documents and that the contractor should be kept informed of any changes in these dates.

It is becoming increasingly common for this information to be given to a contractor during a telephone call from a client. This is clearly contrary to the recommendations of the NJCC *Code of Procedure for Single Stage Selective Tendering*, but in recognition of reality a checklist form (see Figure 2.1) should be available to those taking the initial enquiries.

The initial decision to tender is made by the chief estimator, often in consultation with general management, using the data recorded on the Preliminary Enquiry form. The decision process is dealt with in detail in the next chapter.

2.6   TENDERING QUESTIONNAIRE

Preselection data are usually obtained by means of a questionnaire sent by the client or consultant to various contractors.

The 1994 report by Sir Michael Latham *Constructing the Team*, comments on the way in which many public bodies locally and nationally maintain lists of suitable contractors (and consultants). With so many questionnaires used to pre-qualify for projects there is an enormous '...duplication of effort and wasteful burden for the construction industry'. The report recommends a single qualification document which could be accessed by any public sector body. Clients should consider a shorter form to reduce the burden on smaller contractors for low value contracts.

The National Joint Consultative Committee for Building (NJCC) has published a Standard Form of Tendering Questionnaire so that contractors can prepare answers in advance.

In addition to providing essential project information to a contractor, clients and their consultants may well find it advantageous to produce a standard questionnaire for completion by contractors in order to obtain comparable relevant information concerning the resources, management expertise and experience. Such a questionnaire should seek to obtain the following information from the contractor:

- Name of firm.

- Registered office address.

Figure 2.1 Preliminary enquiry form

PRELIMINARY ENQUIRY FORM	KEYSTONE CONSTRUCTION LTD

Contact

Name	
Organisation	
Telephone	Fax
Comments	

Project details

Brief description

Location

Client name

Consultants

Conditions of contract

Documentation

Programme

Tender details

Tender period	Document issue	Tender date
Estimated value		
Number of tenderers		
DECISION	**Accept**	
	Decline	
Signed	Date	

- Local office address and telephone number.

- Date of formation or registration.

- Registration number.

- If a member of a group of companies, the name and address of the parent company.

- Whether the parent company guarantees the performance of its subsidiary in accordance with the proposed terms of contract.

- Nominal and paid up share capital.

- Particulars of insurances held by the contractor.

- Annual turnover during the past three years and current workload.

- Details of contracts of a similar nature and size in hand, or carried out during the last three years, including:

  — description of project

  — client

  — architect

  — quantity surveyor

  — engineer

  — value

  — contract period and information on actual completion date compared with programmed completion date

- Number of employees employed on a regular basis under the following headings:

  — administrative

  — technical

  — operatives

- List of trades usually sub-contracted.

- Name and addresses of referees from whom references may be obtained.

- Names and qualifications of the directors and the proposed management team for the project.

- Other comments that may be of assistance to the client in arriving at a selection decision, including method of financial and programme control, organisation structure and preliminary ideas on method of construction.

- The person dealing with the application in the contractor's organisation.

Clients will wish to be assured of the financial stability of the contractors who will be invited to pre-qualify. Most contractors will provide bank or trade references if required. It should be borne in mind by those inexperienced in building that balance sheets and financial reports of construction companies can be complex and, to the uninitiated, misleading. Care should be taken in interpreting such financial information.

For large and complex schemes, contractors and trade specialists are invited to examine the methods, resources and timing that would be applied to the project. In a very short time contractors must produce outline programmes and method statements often with little more than a one-paragraph description of a scheme.

2.7 CONTRACTOR'S RESPONSE, PRESENTATION AND INTERVIEW

Contractor's response

On receipt of an enquiry requesting a contractor to pre-qualify for a project, a basic record form, such as Figure 2.1, must be completed. This form will summarise for management the basic data concerning the project. At this stage all of the information may not be available but there are advantages in recording the data on a form which will also be used when a firm enquiry to tender is received. This will allow the sequential completion of the form as information is available. The 'Tender information form' shown in Figure 3.1 (see page 19) will be completed later when more detailed data is available.

The task of completing the preselection questionnaire and providing supplementary data to support any request for pre-qualification often falls to the estimating department. As a consequence, management must ensure that the estimating department remains fully informed of the company's workload objectives, projects in hand and completed, and that it has access to the required financial and personnel data concerning the company that is sought from preselection questionnaires.

The contractor's response to an initial enquiry is based on the recommendations of management in conjunction with the estimating department on the following points:

- Suitability of the project with regard to the company's future workload requirements.

- Financial resources and commitments of the client, exhibited by proven financial backing and, for development work, sufficient tenant agreements signed for the project to start.

- Market conditions, including the general economic and political situation and outlook.

- Construction problems.

- Previous experience with similar types of project.

- Previous experience with the client or consultant.

- Adequacy and quality of information provided.

- Resources needed to carry out the project in accordance with the client's requirements.

- Workload of the estimating department.

- Seasonal weather risks.

- Risks imposed in the conditions of contract.

- Risks in the construction of the project.

- The number of contractors to be invited to tender.

Presentation document

It is common for contractors to prepare a written presentation document for each proposed project. This contains high-quality charts and illustrations to show their organisation in the best light. Large companies have a proposal production department which is used at all stages of marketing and scheme development.

The topics which are common to most proposal documents are listed below and can be used as (off-the-shelf) core themes:

- Introduction to the company and mission statement.

- Areas of operation and projects undertaken.

- Resources:

    — managerial, supervisory and technical staff

    — equipment

    — links with trade specialists

- Quality assurance procedures.

- Safety management.

- Photographs of similar or prestige projects already built.

- Lists of other clients.

- Statements on training, community liaison, equal opportunities, and environmental issues.

Interview

On larger projects it may be necessary for the contractor to meet with the client and consultants for interview. This allows both parties to elaborate on their proposals and to discuss the project and method of working in more detail. It is important at this stage that (where possible) the contractor's personnel who will be considered for the project should be present, as well as those who will prepare the tender. Also, the consultants directly responsible for the project should attend, as well as senior representatives of the practices involved in the project. An early opportunity for key personalities associated with the future project to meet is advantageous to all.

Before contractors are invited to attend an interview, an agenda must be written enabling those in attendance to prepare appropriately for the meeting. A typical structure for this meeting is as follows:

1.0 Introductions

2.0 Scheme outline and objectives

3.0 Contractor's presentation

- approach to managing the project
- resources
- methods
- planning
- services management

4.0 Questions

5.0 Contractor's summing-up

6.0 Client's summing-up and timetable

Contractors should ensure that they know the time available for their presentation. A team leader must be appointed and a rehearsal is essential. The following guidelines will enhance the quality of a presentation:

- Preparation

    — Obtain an agenda from the client and the names/roles of the interviewers.

    — Establish the aims of the presentation. '

    — Agree which topics need to be covered.

    — Write a rough draft of key points for each topic.

    — Structure the presentation to accord with the agenda.

    — Consider who will speak on each topic.

    — Introduce visual aids to hold the attention of the audience.

- Rehearsal

    — The team should meet shortly before the formal presentation.

    — The team leader can ask typical questions.

— Introduce enthusiasm to the subjects.

— Emphasise co-operation, avoid contention and talking down to the client or his consultants.

Although clients will want information about previous successful projects, it is important to concentrate on the scheme under consideration and investigate ways to identify and solve technical and commercial problems.

2.8 TENDER LIST AND INVITATION TO TENDER

The response from the contractor, if positive, will indicate to the client and his consultants whether the contractor has the necessary expertise, experience, management, resources (including financial) and knowledge of the location to adequately construct the works in the time required. Adequate and consistent appraisal of contractors submitting information is essential.

From the written exchange of information, and any interviews felt necessary by the client and his consultants, a short list of contractors will be produced. This preselection procedure will produce a list of companies, any of which would be capable of constructing the works.

Consultants should advise contractors who are selected at this preselection stage that they will be receiving tender documents on the advised date. The contractor can then reserve appropriate resources to process the tender when it arrives. It is important that these dates are adhered to and that the contractor is advised of any changes that occur. Contractors who are not selected at the preselection stage should also be advised promptly that they have not been successful.

Once a contractor has indicated his willingness to submit a tender, he should only in exceptional circumstances, subsequently withdraw from submission of a tender. If a situation occurs which does cause the need for such a withdrawal, such as an unexpected increase in workload which commits the contractor's resources, then maximum notice should be given to the client or consultants of such withdrawal.

3. DECISION TO TENDER

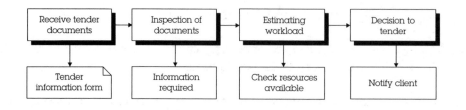

3.1 INTRODUCTION

The decision to tender can occur at one of two stages:

- When preselection enquiries are initiated by a client or his consultants, the contractor will make a decision based upon an outline of the tender information available at that stage. This intention to submit a tender must be re-affirmed when the full invitation to tender and supporting documentation are received.

- When the preselection procedure has not been followed the contractor may find that tenders arrive without prior notice. In such instances, only one opportunity exists to appraise the project and decide whether or not to tender for a project.

In the case of a project where preselection has occurred and details have already been sent to the contractor, a checking procedure is needed to confirm that the project conforms with the information already provided and that the contractor's position regarding tendering has not changed. It is essential that adequate time is allowed for this procedure by clients and consultants. In either situation, the contractor will follow the same procedures in analysing the information received.

3.2 RECEIPT OF TENDER DOCUMENTS

On receipt of tender documents a 'Tender information form' (Figure 3.1) must be completed by the estimating department and an acknowledgement of receipt of the enquiry sent to the client or his consultant. This form will provide management with a summary of the project and the tender documentation, it will be circulated to managers in other departments and will be a significant document contributing to the decision for the submission of a tender.

Where an approximate cost of the project is provided this must be reconciled against any advice given at the time of preselection. If no

Figure 3.1 Tender information form

KEYSTONE CONSTRUCTION LTD	**TENDER INFORMATION FORM**	Tender number

Project particulars

Project title	Client
Project address	Architect
Project description	
Planning supervisor	Engineer
Contact for site visit	Quantity surveyor
Drawings available at	

Tender information

Date enquiry received		Estimated cost breakdown	£
Date for tender submission		Own work	
Tender validity period		Domestic sub-contractors	
Type of tender Open ☐ Negotiated ☐		Project overheads	
Selective ☐ Other ☐		PC and provisional sums	
Documents BOQ ☐ No BOQ ☐		Total	
Documents to be submitted		Source of budget	

Contract details

Form of contract	Method of measurement	
	Period of interim certificates	
Amendments	Payment intervals	
	Period for honouring certificates	
Deed	Retention	
	Defects liability period	
Insurances	Liquidated damages	
	Fixed price ☐ Fluctuating price ☐	
Bonds/Warranties	Programme Start	
	Duration	
Notes		
Distribution		

approximate cost is given, an early assessment must be made by the estimator to determine the approximate cost of the project and the scope of the works.

3.3   PROJECTS FOR WHICH PRESELECTION HAS NOT OCCURRED

When the contractor receives tender documents for a project for which preselection has not occurred, there is increased urgency in deciding whether to tender. The documents may be unexpected and cause problems in an estimating department which already has a full workload. The project may be of particular interest to the contractor and so it is vital that an early appraisal is made of the potential of any new enquiry.

If the project cannot be accommodated because of workload in the estimating department, or the company's workload needs, then the client must be advised, at the earliest opportunity, that a tender will not be submitted. This will allow time for the selection of another contractor by the client or consultant if it is considered necessary to retain a full tender list.

If the initial assessment of the tender documents indicates that the project is of interest to the contractor, and if estimating department workload commitments allow, then a full inspection of the tender documents will be necessary.

3.4   PROJECTS WHERE PRESELECTION HAS OCCURRED

The first operation will be to check that the tender documentation agrees with the information given at the time of preselection. If this initial inspection discloses discrepancies in the project information, programme, conditions of contract or other areas, the consultants should be advised accordingly. The contractor will then need to examine the tender documents in detail to establish whether the changes are of such significance to discourage tendering.

If the initial inspection confirms that the project conforms to the information given at the time of preselection, the contractor will now proceed with a detailed examination of the tender documents.

3.5   INSPECTION OF THE TENDER DOCUMENTS

The inspection of the tender documents must be made by the estimator responsible for the production of the cost estimate, and a checklist must be established of the documents received. In larger organisations, the documents may be inspected by the chief estimator and other members of the contractor's organisation, including the tender planner, quantity surveyor, buyer, contracts adviser and contracts manager.

Clear lines of communication are needed to ensure that all viewpoints of those examining the documents are considered. The estimator will be responsible for the co-ordination of these views.

Inspection of the tender documents must seek to achieve the following objectives:

- The documents received are those in the enquiry.

- The documents and information are adequate for assessing costs.

- Sufficient time is available for production of the tender.

3.6   INFORMATION REQUIRED FOR PREPARATION OF AN ESTIMATE

For the preparation of a tender programme, estimate and tender, all the information contained in Chapter 2 will be required, together with additional information. This is supplied in various forms, including:

- Drawings.

- Specifications (including performance specifications where appropriate).

- Schedules.

- Technical reports.

- Programme work periods for major nominated sub-contractors.

- Bills of quantity.

All documents must show the date of original issue and the date and nature of subsequent amendments. The drawings listed in the bills of quantity or drawing issue forms will be issued with the tender documents.

Site layout drawings

Drawings must show the existing circumstances of the site and adjoining areas and should include:

- Site boundaries.

- Means of access, including approach roads and entrances, and any restrictions on access.

- Contours.

- Roadways and fences.

- Wooded areas, water and other natural features.

- Overhead, surface and underground lines and equipment of statutory undertakings and other owners.

- Position of buildings on site to be demolished.

- Position and height of buildings on site to be retained.

- Position and height of structures adjoining the site.

- Position and depth of existing foundations and groundworks.

- Temporary structures.

- Outline of new buildings and site works.

- Working areas.

- Restricted areas.

- Location of strata investigations.

- Existing services on or near the site.

- Block plan showing roads, natural features and site orientation, sufficient to permit the location of the site to be ascertained.

Other drawings

- General arrangements for the project, shown by plans, sections and elevations.

- Works to adjoining structure.

- Details of items requiring mechanical lifting, including size, weight, location and fixing needs.

- In the case of alteration work, surveys of the existing structure.

- Specifically designed structural temporary works.

- Special risks, construction methods proposed and sequence related to engineering considerations.

Detailed drawings, outline drawings and schedules

Detailed drawings will be required to give information sufficient to determine the location, nature and dimensions of work to:

- Sub-structure.

- Frame.

- Upper floors.

- Roof.

- Canopies, balconies and projections.

- Staircases.

- External cladding.

- Non-standard joinery.

- Purpose-made doors and metal windows.

- Structural walls.

- Non-structural walls.

- Suspended ceilings.

- Finishes.

- Drainage internal.

- Drainage external.

- External work.

- Builder's work in connection with services.

Outline drawings will be required to give information sufficient to determine the location and nature of work to:

- Waste, soil and overflow pipes.

- Cold water services.

- Hot water services.

- Heating services.

- Ventilation services.

- Gas services.

- Electrical services.

- Other services.

Outline drawings or schedules will be required for work to:

- Partitions.

- Fittings.

- Sanitary fittings.

- Non-standard joinery.

Schedules may be required (if not available elsewhere) for:

- Standard joinery.

- Standard windows, frames and ironmongery.

- Standard doors, frames and ironmongery.

- Miscellaneous ironmongery.

- External wall finishes.

- Internal wall finishes.

- Floor finishes.

- Ceiling finishes.

- Decorations.

Technical reports

The following technical reports will be required:

- Site investigation report, including water table and an interpretation of factual data. This information is needed for piling, excavation methods, disposal of surplus excavated material and to consider the effect of chemicals on permanent works.

- CDM health and safety plan which identifies hazards likely to be encountered during construction, stating where and when they are likely to occur. The risk of a particular hazard occurring must be assessed by the planning supervisor. Before tendering, the safety plan must be sufficiently developed for it to form part of the contractors' tenders.

- Special site conditions such as fire risks, security risks or radiation hazards.

- Other technical reports, such as the condition of an existing structure or special requirements for temporary works or plant.

- Reports on hazardous materials such as asbestos, contaminated ground and the need for decontamination of buildings.

Special requirement of the client

- Programme and phasing requirements.

- Security conditions for sensitive sites; either in relation to the site or its surroundings.

- Provision of drawings of services as installed, for maintenance purposes.

- Training of client's staff in operating services or mechanical equipment.

- Method statement.

3.7 CONDITIONS OF CONTRACT

Many standard forms of contract will be encountered in building. These include: JCT '80, in both local authority and private versions, and with or without quantities alternatives, Intermediate Form tailored for medium-sized building projects needing a less complex contract, GC/Works/1 (latest edition) used for central government projects, the Agreement for Minor Building Works used on smaller jobs, and other more specialist forms used by larger public sector and private companies.

The particular contract conditions must be identified and any changes to standard forms of contract noted. Minor alterations can significantly affect the contractor's cash flow, the funding needs of the project, the risk allocation and the responsibilities of the respective parties to the contract. Such changes must be highlighted by the estimator – or contracts specialist – and information passed to management on the extent of the inherent risk to the contractor of any changes made. Such advice will be of significance in any decision to tender.

Supplementary conditions are often attached to contracts dealing with insurances, warranties, design responsibilities and performance bonds. Where onerous non-standard conditions are attached to a contract many contractors either ask to be excused from tendering or add a premium to the tender bid.

3.8 WORKLOAD AND TIMETABLE

Two separate decisions must be made concerning workload. The chief estimator must make the decision concerning the workload of the estimating department, and be satisfied that the estimator who will be allocated to a project has the necessary expertise and knowledge needed for that particular estimate. The advantages of preplanning, which will be possible from advance warning by preselection, will be obvious.

The second decision must be made by company management. This concerns the objectives and needs of the company for workload and also the availability of resources to construct the project. Management must be satisfied that the project meets the company's objectives regarding workload and that the company is not exposed to undue risk by commitment to excessive amounts of work with one client, one project or one sector of industry.

The tender timetable will be of significance to both the estimator and management. The project estimator will normally be responsible for managing the production of the estimate and it will be essential for consultants to allow sufficient time for the assimilation of the project information, obtaining quotations from trade specialists and the production of the estimate. The result of inadequate time may be that a contractor will decline an invitation to tender or the most economic solution will not be produced.

3.9   TYPE OF WORK AND RESOURCES NEEDED

In making a decision to tender, the contractor will initially examine the type of work involved in the proposed project. Many contractors concentrate on particular market sectors, such as new build or refurbishment, industrial, commercial or housing, while others prefer to maintain a spread of work across various building types. Whatever the corporate strategy of a company, an objective decision must be made as to whether a new project fits into the company's workload objectives at that time. Such a decision must also consider the resources currently available to the company.

Resources include finance, staff and labour – bearing in mind the particular skills and quality needed for a project, as well as the availability of materials and plant. The estimator contributes to this assessment with his knowledge of market conditions, trends in the costs of basic resources and in the identification of problem or risk areas.

The financial appraisal involves an assessment of the payment proposals in the conditions, any bonding arrangements, the relative proportions of work to be carried out directly by the contractor, sublet to domestic sub-contractors or carried out by nominated sub-contractors. From these assessments, the funding requirements of the contract will be established and finance charges can be calculated.

Decisions in these matters must be taken by senior management. Careful definition of company objectives and policy will allow chief estimators to concentrate the resources of the estimating department on the most suitable projects for the company when the enquiry is received.

3.10 TENDER INFORMATION

The estimator is particularly concerned with the quality of the information provided at the enquiry stage. Inadequate or contradictory information leads to a cautious approach by the estimator. While the estimator should seek further information from the consultants in such cases, the estimator's approach is often coloured when uncertainties exist and the project is not fully defined. Past experience may well indicate that the uncertainties continue into the construction phase, with dire consequences for both the project and the contractor.

Bills of quantity were once fully descriptive of the works in themselves, without the need for reference to other project information. With the introduction of *Co-ordinated Project Information* (CPI), and the widespread use of the *Standard Method of Measurement for Building Work* (SMM7) and the *National Building Specification* (NBS), bill descriptions are shorter. The effect is that the contractor needs to provide more resources for the preparation of an estimate.

The project specification is now clearly linked to work sections by a standard coding system. Any deviations from the Standard Method of Measurement must be noted. Where bills of quantity are not provided, good drawings together with a comprehensive specification will be required, as described in Section 3.6.

The quality of design and evidence of co-ordination of services will be of significance to the estimator. The extent of design advancement, as reflected in the drawings provided, must be compared with the bills of quantity to ensure that the bills accurately reflect the fully designed work. Note must be taken of provisional and prime cost (PC) sums where design is not yet complete.

The estimator must obtain from the project information – usually the preliminaries clauses – details which:

- Relate to the contractor's intended method of working.

- Impose restrictions.

- Affect access to the site.

- Interrupt the regular flow of trades.

- Affect the duration of the project.

- Require specialist skills or materials.

- Have a significant effect on the programme.

- Are of major cost significance.

Such items affect costs and will guide the estimator during the production of the estimate. It is also essential that phasing requirements are clearly defined not only in the contract documents but also by illustration on drawings. Inadequate definition of such items will lead to problems concerning programming and definition of cost.

The estimator must co-ordinate the views of the various members of the contractor's team who examine the project information; and must provide management with a realistic appraisal of this information to enable a decision to be made on whether or not to tender for the project.

3.11 MANAGEMENT'S DECISION TO TENDER

Comments on the contract conditions, workload, type of work and resources needed and the tender documentation will all be taken into account by management in deciding whether a tender will be submitted. The cost of preparing a tender must be viewed in relation to the chance of being awarded the contract.

If it is decided to accept the invitation, the procedure for the preparation of an estimate described in the following chapters of this Code should be commenced and the consultant informed that a tender will be submitted. If it is decided that the invitation to tender will be declined, a verbal response should be given and all documents should then be returned immediately to the consultant with the reason why a tender will not be submitted. This action, if taken, must be carried out as soon as possible after receipt of the tender documents.

In either case, the 'Tender information form' (Figure 3.1) should be filed in a register with the decision noted and circulated to departmental managers.

4. PROJECT APPRECIATION

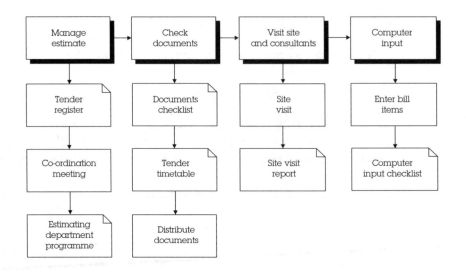

4.1 MANAGEMENT OF ESTIMATES

Project appreciation commences in the 'Decision to tender' phase but intensifies once management has confirmed that an estimate is to be prepared. All projects in an estimating department need to be entered in a tender register (an example is given in Figure 4.1). The tender register is kept by the chief estimator in a secure place for reference over a number of years. Many companies use the register to evaluate tender performance in relation to their competitors.

Considerable management skills and personal leadership will be needed by the estimator to motivate and co-ordinate the various staff associated with the preparation of the estimate and its subsequent conversion into a tender.

Co-ordination meetings are needed with management and other departments within the contractor's organisation to establish key dates, decide on actions necessary and monitor progress during the production of the cost estimate. An example of a typical checklist for such co-ordination is given below:

1.0 Project location and description

2.0 Parties

3.0 Programme – latest dates for:
- despatch of enquiries
- receipt of quotations
- visit site

Figure 4.1 Tender register

KEYSTONE CONSTRUCTION LTD	**TENDER REGISTER**	Start _____ Finish _____

Tender number	Client	Title	Tender		Result		Notes
			Date	Sum	Yes % low	No % high	
10166							
10167							
10168							
10169							
10170							
10171							
10172							
10173							
10174							
10175							
10176							
10177							
10178							
10179							
10180							
10181							
10182							
10183							
10184							
10185							
10186							
10187							
10188							
10189							
10190							
10191							
10192							
10193							
10194							
10195							
10196							
10197							
10198							
10199							
10200							

- complete tender programme
- attend co-ordination meetings
- complete estimate
- price project overheads
- final review meeting

4.0 Work packages and suppliers

5.0 Commercial risks, responsibilities and opportunities

6.0 Distribution of documents

4.2 ESTIMATING DEPARTMENT PROGRAMME

The chief estimator will prepare a bar chart (see Figure 4.2) each week to plan the estimators' workload showing both present and possible future tenders. Copies are sent to heads of other departments so they can establish their input.

4.3 CHECK TENDER DOCUMENTS RECEIVED

Once the decision has been taken to tender for a project, the estimator must ensure that all the tender documents have been received. A check must be made to see that all drawings received are of the revision noted, and that all other documents listed in the invitation letter are provided. Figure 4.3 shows a checklist for tender documents which will help the estimator to assess their suitability for the estimate. A letter is sent to the client or consultant to acknowledge the receipt of the documents and confirm that a tender will be produced by the due date. This letter should also record any discrepancies in the documents received.

Drawing lists can be checked against drawing issue sheets which usually accompany the documents. It is not necessary to produce issue sheets for suppliers and sub-contractors because their documents are recorded on enquiry abstracts (see Figures 5.1 and 5.2 on pages 49 and 50), and listed in enquiry letters. A record of any subsequent additional or revised documents should also be carefully recorded in the estimator's file.

It is important to make an accurate record of tender documents received because they will become the basis of a formal offer and

KEYSTONE CONSTRUCTION LTD		ESTIMATING DEPARTMENT PROGRAMME				From
Tender number	**Project title**	**Approximate value**	**Type**	**Estimator**	**Tender date**	
10166	Hamilton Housing, Hamfield	2,000,000	D&S	JG		
10167	Pickson Packing Factory, Newport	2,200,000	D&B	JP		
10168	Omega Industrial Estate, Fencing	150,000	D&S	CB		
10169	Campton Primary School	1,850,000	BOQ	BB		
	Dock access road	280,000	BOQ			
	Market re-roofing, Banton	750,000	D&S			
	Hospital extension, Harpley	1,250,000	BOQ			

Key

Present tenders		Drawings and specification	D&S
Future tenders		Bills of quantity	BOQ
		Design and build	D&B

Circulation

Date

Figure 4.2 Estimating department programme

Figure 4.3 Tender documents checklist

KEYSTONE CONSTRUCTION LTD	TENDER DOCUMENTS CHECKLIST		Tender number					
			Distribution					
						Design & build		
	Date received	Number of copies	Planning	Commercial	Buying	Architect	Engineer	QS
Formal tender documents								
Invitation to tender								
Tender instructions								
Form of tender								
Return envelope								
Drawings								
Architectural								
Engineering								
Services								
Schedules								
Specifications								
Building								
Engineering								
Services								
Bills of quantity								
Building								
Engineering								
Services								
Schedule of works								
Schedule of rates								
Project specific								
General								
Conditions of contract								
Articles of agreement								
Appendix details								
Payment schedule								
Programme								
Design liabilities								
Parent company guarantee								
Bond								
Warranties								
Technical reports								
Health and safety plan								
Method statements								
Site investigation								

eventually be checked against the contract documents in the case of successful tenders. This can best be achieved by stamping each drawing and document to show the date received and the words 'Tender document'. This procedure will later help to indicate their purpose and to differentiate them from construction issues once a contract has been awarded. Where drawings are copied for sub-contractors and suppliers, they should be stamped to avoid confusion between 'tender' and 'construction' copies.

4.4 TIMETABLE FOR PRODUCTION OF ESTIMATE AND TENDER

The estimator must, as the manager responsible for the production of the estimate, ensure that a timetable is established which highlights the key dates in the production of the estimate and tender (see Figure 4.4). It will be an essential document for all those associated with the tendering function. The following dates must be established without ambiguity:

- Latest date for despatch of enquiries for materials, plant and sub-contracted items.

- Latest date for the receipt of quotations.

- Bills of quantity production for design and build, drawings and specification contracts.

- Visit to the site and the locality.

- Finalization of the method statement.

- Completion of pricing and measured rates.

- Finalization of the tender programme.

- Intermediate co-ordination meetings within the contractor's organisation.

- Review meetings.

- Submission of the tender.

All personnel associated with the tender must confirm that they are able to provide the necessary data in the format required, in accordance with the agreed timetable.

TENDER TIMETABLE

Project
Helix Laboratories, Westfield

KEYSTONE CONSTRUCTION LTD

Description	Month	April														May																			
	Date	11	12	13	14	15	18	19	20	21	22	25	26	27	28	29	2	3	4	5	6	9	10	11	12	13	16	17	18	19	20
	Latest date	M	T	W	T	F	M	T	W	T	F	M	T	W	T	F	M	T	W	T	F	M	T	W	T	F	M	T	W	T	F

Project appraisal
Check documents
Tender information sheet
Tender timetable
Document production [d & b and drg & spec]
[Drawings]
[Specification]
[Bills of quantities]
Code bill items and enter computer bill
Enquiries
Abstract, prepare, despatch – subs — 18 Apr
Abstract, prepare, despatch – mats — 20 Apr
Date for receipt of quotations – mats — 4 May
Date for receipt of quotations – subs — 9 May
Project appraisal
Site visit — 19 Apr
Tender method statements
Tender programme — 12 May
Pricing
Labour and plant
Materials
Sub-contractors
PC and provisional sums
Project overheads
Reports
Checking procedures and summaries
Tender
Review meeting(s) — 17 May
Submission documents
Submission — 19 May

Figure 4.4 Tender timetable

4.5   THOROUGH EXAMINATION OF TENDER DOCUMENTS

The conditions of contract, bills of quantity, general arrangement drawings, specification and any supporting documents, such as soil reports, site industrial relations policy or specialist information must be examined in detail by the estimator.

At this stage, the tender documents may also be examined by other personnel in the contractor's organisation, including those associated with purchasing, tender planning, commercial management, engineering, temporary works, plant and construction. The estimator must maintain records of those personnel who receive copies of the tender documents. The form shown in Figure 4.3 is used for this purpose. Any discrepancies or divergence between any of the tender documents must be noted by the estimator for future reference to the appropriate consultant for clarification.

Conditions of contract and appendix

Any unusual features in the conditions of contract and appendix will have been identified prior to the decision to tender and noted on the 'Tender information form'. At this stage the estimator must consider in more depth the consequences of such conditions and note on the 'Estimator's summary' his recommendations for dealing with the situation (see Figure 12.1 on page 173).

Any further particulars received during the tender period concerning the conditions of contract and appendix details, must be noted and entered in the 'Estimator's summary' and the estimator must ensure that appropriate action is taken regarding any cost implications of revised or additional information received.

The examination of conditions of contract and appendix must reveal whether:

- Non-standard forms of contract are used.

- Conditions of standard forms are amended.

- Non-standard payment or retention provisions are included and the affects that such alterations may cause regarding the contractor's liability for payments to subcontractors.

- Bid bonds, performance bonds or parent company guarantee are required.

- Insurance requirements are met by the contractor's standard policies, or can be obtained within the current insurance market at reasonable rates.

- Nomination procedures will be followed without amendment.

- Liquidated damages requirements are acceptable and whether any allowance should be made in the tender.

- Amendments to other project information have been reflected in the condition of contract, i.e. items not measured in the bills of quantity in accordance with the SMM.

Specification, bills of quantity and drawings

The estimator must have a thorough understanding of the specification. This document must be examined in detail and compared with the items measured in the bills of quantity and tender drawings, to identify any divergence and discrepancies.

With the widespread use of SMM7 and the National Building Specification (NBS) bill descriptions are shorter. Reference is made within bill items to specification clauses using the Common Arrangement of Work Sections (CAWS). The estimator must ensure that there is consistency in use and that the cross-references are accurate. Some clauses can make reference to other clauses within a specification which can lead to confusion. If, for example, an enquiry to a sub-contractor for patent glazing includes the appropriate clause H74, the estimator will need to consider other linked clauses such as those for glazing and sealant. Furthermore, an amendment during the tender period to a clause can be very difficult to allocate to the various subcontractors who might be affected.

The specification must be clear and unambiguous and the estimator must prepare a queries list of any unclear items for discussion with the appropriate consultant.

A schedule must be prepared for use in obtaining quotations for various materials and elements to be sub-contracted. In order to prepare this the estimator must first of all group together relevant items from the specification and bills of quantity into respective trades and, if necessary, individual suppliers. It may be necessary to obtain further information from drawings, for example, lengths of timber may need to be specified. The estimator should consider any additional quotations that will be required to enable him to have flexibility when pricing. This could include extra-over rates for machine off-loading for bricks, for example, or extra charges made by ready-mixed concrete suppliers. Such information will be of benefit when pricing the bills in detail at a later stage. (A sample sheet for abstracting specification and bill of quantity items for materials and sub-contractors enquiries is shown in Figures 5.1 and 5.2, enquiry abstract sheets for materials and sub-contractors (see pages 49 and 50).

The bills of quantity must be carefully analysed and any divergence from the SMM noted and drawn to the attention of the consultants, particularly when an estimator is concerned that poor specifications could make his tender less competitive.

Attention must be paid to any small quantities of similar materials which are noted within several sections of the bills of quantity. When aggregated together, such items may form a significant total, but their use on site may require a series of small deliveries at premium rates. Notes should be prepared of such requirements for use when enquiries are sent out and when unit rates are being prepared.

The information required on tender drawings has already been described in Section 4.3 and will be a significant factor in the decision to tender. At this stage, the estimator will systematically analyse the data shown on the various tender drawings and schedules and build up a picture of the project's needs. This analysis must highlight the cost-significant items in the project and demonstrate areas of caution or risk. Further information may be needed when the site is visited. This information must be recorded for review purposes in the 'Estimator's summary'.

At this stage, the estimator should look for various factors which will influence his approach to pricing, such as:

- Standard and completeness of the drawn information.

- Tolerances required.

- Clarity of the specification requirements and the quality required.

- Buildability.

- Whether load-bearing and non-load-bearing areas can be identified.

- The extent of use of standard details indicating previous construction experience.

- Evidence of design co-ordination of services and structural needs.

- The amount of information concerning ground conditions and foundations.

- Problem areas and restraints on construction in the design.

From this appreciation of the drawn information, the specification and measured items in the bills of quantity, the estimator will begin to understand how the job is to be built and the quantity and quality of resources needed. It is unsatisfactory that while designers often take several months or even years to arrive at this stage on a project, the estimator and his team will be required to assimilate all the information produced, decide on how the project will be built and estimate its cost of construction, often within a tender period of a few weeks.

Outstanding information

The thorough examination of the tender documents by the estimator will highlight queries concerning the project and the method of construction. It is essential that the estimator lists these and collates together queries that are raised by other members of the contractor's team who have examined the documents. It is important that all queries concerning the tender information are channelled to the estimator for resolution. Ease of communication and good management practice dictate that only the estimator should deal with the consultants at this stage.

Queries will be resolved by meetings and discussions between the estimator and other members of the contractor's team and the consultants or the site visit. The estimator must ensure that all queries raised with the consultant are made in writing and that the urgency of the information needed is stressed. Any significant decisions communicated to one contractor by the consultant, must be communicated to all tendering contractors.

4.6   DISTRIBUTION OF DOCUMENTS

The estimator has a responsibility to ensure that other members of the contractor's estimating team with specialist knowledge or interests are provided with copies of all relevant project information.

The tender documents checklist shown in Figure 4.3 can be used to record the distribution of documents. In larger organisations specialist departments will assess the risks and obligations as follows:

- An insurance manager will obtain a quotation for a specific project if it exceeds a certain value or involves working in a hazardous location.

- A legal department or company secretary may need to comment on the wording of warranties and performance bonds or arrange directors' signatures for parent company guarantees.

- The commercial manager, or quantity surveyor, is often consulted for advice on the proposed conditions of contract, in particular the terms of payment which are critical to the financial success of a contract.

4.7 VISITS TO SITE AND CONSULTANTS

Site visit

The visit to the site must be made once preliminary assessment of the project has been carried out and a provisional method and sequence of construction established. The estimator may be accompanied on such a visit by other members of the contractor's team.

As well as visiting the site itself, the opportunity must be used to examine the general locality and to establish the extent of other building works in the area. Visits should also be made to local labour agencies and suppliers in the area. Any opportunity to visit excavations near to the site must not be missed and, in this connection, the local authorities may be able to give advice on local conditions and of any excavations which may be visible adjacent or near to the site.

A comprehensive report of the site visit must be prepared, and a sample form is shown in Figure 4.5. Site photographs can often be a useful way of recording information for discussion and record purposes.

Points to be noted when making a site visit include:

- Position of the site in relation to road and rail and other public transport facilities.

Figure 4.5 Site visit report

KEYSTONE CONSTRUCTION LTD	SITE VISIT REPORT	Tender number
		PAGE 1 OF 2

Project particulars	
Project title	**Contact for site visit**
	Name
Project address	Tel. number
	Fax number
	Visit details
Directions to site	Person making visit
	Date Time

	Remarks
Site position	
In relation to public transport	
Distance from local office	
Other work in the area	
Adjacent buildings	
Fencing and hoardings	
Demolition	
Hazards	
Site conditions	
Topography	
Trees and vegetation	
Site clearance	
Ground conditions Borehole details	
Type of soil	
Stability of soil	
Water table	
Tidal conditions	
Pumping requirements	
Disposal of water	
Security problems	
Weather exposure	
Space for temporary accommodation	
Restraints for static plant Cranes	
Scaffolding	
Live services	
Protection requirements	

KEYSTONE CONSTRUCTION LTD	SITE VISIT REPORT	Tender number
		PAGE 2 OF 2

	Remarks
Access	
Temporary roads	
Safety	
Deliveries	
Traffic restrictions	
Local facilities	
Disposal of soil	
Services Telephone	
Water	
Electricity	
Sewers	
Garages, refreshments	
Local authorities	
Statutory authorities	
Local contacts	
Security services	
Labour agencies	
Plant hire	
Local tip and charges	
Sub-contractors	
Suppliers	
Sketches/photographs attached	

- Names and addresses of local and statutory authorities.

- Topographical details of the site, including note of trees and site clearance required.

- Any demolition work or temporary work needed to adjacent buildings.

- Access points to the site and any restraints on layouts that have been considered.

- Ground conditions: any evidence of surface water or excavations indicating ground conditions and water table.

- Facilities in the area for the disposal of spoil.

- Existing services, water, sewers, electricity, overhead cables, etc.

- Any security problems, evidence of vandalism, need for hoardings, lighting, etc.

- Labour situation in the area.

- Availability of materials.

- Weather conditions – high rainfall, winds, etc.

- Temporary roads.

- Location of nearest garages, hospital, police and cafés.

- Nature and use of adjacent buildings, such as industrial or residential.

- Police regulations.

- Local sub-contractors.

- Restraints imposed by adjacent buildings and services, i.e. space available for tower cranes, overhang, etc.

- Other work currently in the area, or shortly to start.

- Availability of space for site offices, canteen, stores, toilets and storage.

- The effect any client requirements may have upon access, storage, movement or accommodation.

- Special difficulties.

Visit to consultants

The estimator may need to visit the consultants, particularly when further information is needed which has not been given to tenderers, such as further drawings and site investigation reports. In some circumstances these documents may be confidential. Visits will normally be made to the architect, but it may also be advantageous to visit the consulting engineer, services engineer and quantity surveyor, in order to meet the personalities who will subsequently be involved with the project.

Detailed drawings, reports of site investigations and any other available information must be inspected and notes and sketches made of all matters affecting either construction method, temporary works or the likely cost of work.

A critical assessment must be made of the degree of advancement and quality of the design. A well-developed and well-documented design may be indicative of a smooth running and possibly profitable project. A design which is obviously ill-conceived and incomplete may cause delays during construction and allowance should be considered for this in the tender. An adverse report may lead to the reconsideration of the decision to submit a tender.

Visits to consultants have become less common because contractors are usually given copies of all the drawings and specifications which are going to become part of the contract documents. It is more likely that a visit will be seen as an opportunity to show the tenderer's willingness to contribute to the scheme, to work closely with the design team and to express an interest in further work.

4.8   COMPUTER INPUT

The two main uses for computers in estimating are: to keep names of suppliers and sub-contractors for enquiries and to produce a priced bill of quantity.

Much of the estimator's analysis of costs is made using a computer model of a bill, which is entered in a number of ways, as follows:

- Taking-off quantities directly from drawings.

- Using a scanner to transfer written bills of quantity to a text file.

- By manual entry of item references, descriptions (if required), quantities and units.

- Where bills of quantity are available on disk or by electronic mail for immediate incorporation into a computer-aided estimating system.

If a contractor has suitable software, and access to a bill of quantity with full descriptions, enquiries to sub-contractors can be extracted and printed by the computer once the items have been coded by the estimator. For example, all the Piling items may have a sort code S01 and Metalwork S02, which are also the references used for sub-contract enquiries on the sub-contractors' enquiry form shown in Figure 5.2 (see page 50). The computer can then find all items in a trade, print the relevant pages from the bill and take names and addresses from a vendor database.

The checklist shown in Figure 4.6 may be used to ensure that the computer bills accurately represent the written document. The most important check is that all the pricable items are entered and quantities are correct. If an estimator prices the work items directly from the computer (without reference to the written document) then descriptions will be checked, particularly dimensions forming part of the text.

There are many hazards which an estimator must find before passing documents to computer assistants. The bills of quantity may need some simple modifications in order to comply with the estimating software being used. Problems often occur in the following ways:

- Page numbers can be difficult for a computer system to accept, such as: 12A/A/D1 – in this case the estimator could re-number the bill pages with a simpler reference such as 12.1 meaning bill 12 page 1.

- Some bills have work items and collections on the same page; again many computer systems need different page numbers for collections and summaries.

- Any missing item references must be inserted by the estimator.

Figure 4.6 Computer input checklist

KEYSTONE CONSTRUCTION LTD	**COMPUTER INPUT CHECKLIST**	Tender number	
		Date completed	Notes
Estimator			
Check bills of quantity numbering system			
Check integrity of collections and summaries			
Examine preliminaries bill for PC and provisional sums			
Input new project details			
Input PC and provisional sums			
Trade code bill items			
Assistant			
Enter bill items			
Optical character recognition			
ASCII file			
Manual input			
Input collection pages and summaries			
Check quantities			
Estimator/buyer			
Check trade coding			
Select appropriate suppliers and sub-contractors			
Print enqiry letters and bill items			

5. ENQUIRIES AND QUOTATIONS

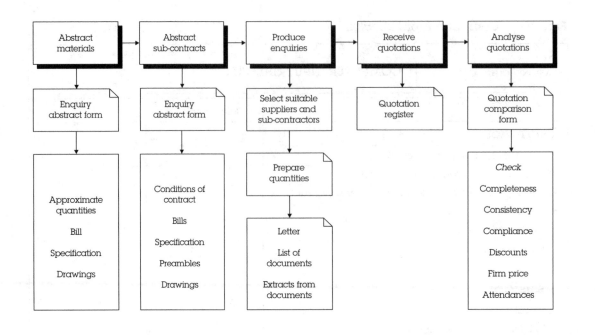

5.1 PREPARATION OF ENQUIRY DOCUMENTS

The contractor's success in obtaining a contract depends upon the quality of the quotations received for materials, plant and items to be sub-contracted. It is essential to obtain realistically competitive prices at the time of preparing the estimate.

The responsibility for carrying out this important function will vary from company to company. In some organisations the estimator prepares the information for enquiry purposes, selects the organisations to receive enquiries and reconciles the quotations received. In others, some or all of this will be done by the buying department, who may provide the estimator with a selection of fully reconciled quotations at the end of the enquiry stage for his final consideration. Standard procedures must be established, setting out the responsibilities of the person who is to carry out the enquiry function and any subsequent negotiations that arise. The professional manner in which enquiries are sent to suppliers and sub-contractors can have considerable effect on the response and quality of quotations submitted.

Lists of items for which quotations are required are established following the detailed examination of the contract documents. Figures 5.1 and 5.2 illustrate typical enquiry abstract forms which enable the estimator or buyer to list the materials and trade packages which need quotations. If only drawings and specifications are provided it is necessary for the contractor to produce additional

KEYSTONE CONSTRUCTION LTD			ENQUIRY ABSTRACT – MATERIALS				Project — Helix Laboratories, Westfield			
							Suppliers			
Ref.	Description	Approx. quantity	Unit	Bill pages	Spec. pages	Drawings	1	2	3	4
M01	Hardcore	6000	m³		D20		Target	Wilson	Dixon	
M02	Type 1 subbase	7200	m³							
M03	Concrete 20 N	250	m³		E10		MBC	Wilson	Montel	
M04	Concrete 30 N	300	m³							
M07	Facing bricks	55.4	th	3/5, 6	F10		Dexter	Opus Trading		
M08	Blocks 140 mm	6550	m²	3/6, 7	F10					
M10	Timber doors	65	nr	3/40, 41, 47	L25	t105	Rosewood	Dale Joinery	Kilroy Timber	
M11	Ironmongery			3/43, 44, 48	P21	t106	Lasermax	Hanson		

Figure 5.1 Enquiry abstract – materials

KEYSTONE CONSTRUCTION LTD	ENQUIRY ABSTRACT – SUB-CONTRACTS					Project Helix Laboratories, Westfield				
Ref.	Description	Approx. quantity	Unit	Bill pages	Spec. pages	Drawings	Sub-contractors			
							1	2	3	4
501	Piling	120	nr	3/1, 2	D31	t101	BBS	Robinson	Dudley Eng	Ross Piling
502	Metalwork			3/36, 37, 39	G12, L31	t110, 111	Dennis Hills	DES Eng	Westfield Eng	
503	Structural steelwork	34	t	3/27, 28	G10	b201, 202	Trafford	King Structures	Steelcare	PCS
504	Roof tiling	1740	m²	3/31, 32	H60	t115, 119	Richard Roofing	AA1 Tiling	Carlton Systems	Lakemore
505	Leadwork			3/32, 33	H71	t119	Richard Roofing	Jan Lead	Carlton Systems	
506	Metal windows	132	nr	3/40, 41	L11	t114, 115	Microtex	Memphis Alum	Columbus Glass	
507	Plastering			3/50, 52	M10, M20	t200	Addford	Riverside Trades	Lister Plastering	RDP
508	Floor coverings	1540	m²	3/55, 56	M12	t200	Floordex	York Flooring	J. Aldridge	

Documents to all   1/2–12
   Health and safety plan
   Site plan t100

Figure 5.2 Enquiry abstract – sub-contracts

information to assist with the enquiries. Every opportunity should be taken to clarify the contractor's requirements, bearing in mind the limited time usually available for obtaining quotations.

The contractor must ensure that comprehensive records are maintained of the various elements of project information sent to suppliers and sub-contractors. These records must list the drawings sent, the relevant contract and specification clauses, project preambles and the pages of bills of quantity.

The use of standardised documentation and procedures assists at this stage in the methodical preparation of the estimate, and allows an interchange of personnel at any stage.

Lists and preselection

Contractors must maintain comprehensive records of suppliers, and sub-contractors. These records must include:

- Details of past performance on site.

- Previous performance in returning complete quotations on time.

- Extent of geographical operation.

- Size and type of contract on which previously used.

- Information concerning contacts.

- Address, telephone and facsimile numbers.

When operating in a new area, a contractor needs information concerning the local suppliers and sub-contractors. In this case performance should be verified from other external sources and any remaining information established from the supplier and sub-contractor concerned.

A questionnaire may be used to establish the resources and abilities of subcontractors concerning:

- Area of operations.

- Size and type of work.

- Labour and supervision available.

- Size and type of work previously carried out.

- References from trade, consultant and banking sources.

- Insurances carried by the sub-contractor (if relevant).

- Confirmation of holding of relevant sub-contractor's tax exception certificate (if relevant).

This questionnaire attempts to establish the supplier or sub-contractor's financial capability to undertake the work in question and to supply the materials and plant required. It is necessary to establish that all resources will be available to meet the requirements of the main contractor's programme.

Preselection will be necessary if dealing with unknown suppliers and sub-contractors. Bearing in mind the particular needs of the project, the contractor must ensure that the list of suppliers and sub-contractors who are invited to tender is comprehensive and that bids will be received. Preselection procedures must confirm that quotations will be submitted and establish that the particular supplier or sub-contractor has the necessary resources and desire to tender for the project. The procedures described in Section 5.2 must be applied if sufficient time is available. In any event confirmation must be obtained from major suppliers and sub-contractors that they will be prepared to submit a bid, before enquiry documents are sent out.

Programme and method of construction

Suppliers and sub-contractors must be advised of the programme requirements and any aspects of the method of construction that are relevant. Programming information must include:

- The anticipated start date for the main contract.

- The approximate start dates for sub-contractor, or materials deliveries.

- The required completion date(s).

- Key information of significance to the progress of the works.

- Phasing.

Suppliers must be given an indication of the rate of delivery, together with any requirement for the approval of samples prior to delivery to site. Although sub-contractors will be provided with details from the contractor's method statement which are relevant to their work, contractors will be reluctant to divulge information which could assist their competitors.

5.2 ENQUIRIES

General

The objective of sending out enquiries is to ensure that technically accurate quotations are received which are compatible with the main contract conditions. Enquiries must set out clearly the terms and conditions on which quotations are being invited. It is recommended that limited numbers of suppliers and sub-contractors are invited to quote for each item or section of the work. This conforms to the principles of selective tendering and avoids the wasteful use of resources in requesting an excessive number of quotations.

The time allowed for the preparation of a tender is invariably short. The contractor has a limited period of time to distribute the relevant enquiries to a sufficient number of suppliers and sub-contractors to ensure an adequate response and yet at the same time limit the number of enquiries to a reasonable number of participants.

Clarity in the enquiry and its supporting documents is essential and every effort must be made to assist the supplier and sub-contractor concerned to avoid:

- The submission of incomplete quotations.

- Failure to meet the required timetable for return of quotations.

- Counter offers of alternatives which require further checking.

Each enquiry should fully describe the work to be undertaken or items to be supplied without ambiguity.

Materials

Enquiries to suppliers of materials must state:

- Title and location of the work, and site address.

- Specification, class and quality of the material.

- Quantity of the material.

- Likely delivery programme, i.e. period during which supplies would be needed with daily or weekly requirements where known. Where small quantities are to be called off from a bulk order this should be clearly stated.

- Means of access, highlighting any limitations or delivery restraints; any traffic restrictions affecting delivery times.

- Special delivery requirements such as palletting or self-unloading transport.

- Date by which the quotation is required.

- Period for which the quotation is to remain open.

- Whether fluctuating or firm price required, the basis for recovery of increased costs and the base date when a formula is used for calculation of fluctuations.

- Discounts required.

- The person in the contractor's organisation responsible for queries.

The contractor has a responsibility to ensure that suppliers:

- Make every effort to meet the specified 'date required'. If this is not possible the contractor must ensure that he is informed promptly so that additional enquiries can be sent out in order to maintain a full enquiry list.

- Clear queries as they arise in order to avoid quotations marked 'more information required'.

- Submit the quotation on time with a clear statement where prices are 'to follow'.

Particular attention should be paid to material supply conditions which may have cost implications. The contractor can incur costs for:

- Pallets left on site.

- Standing time for vehicles while unloading at the site.

- Small quantities or abnormal loads.

- Deliveries at abnormal times.

- Unloading facilities to be provided by the contractor.

- Special height, width or weight of deliveries which might affect the route or access to the site.

- Any escort or special transport arrangements needed.

The contractor must record details of all enquiries sent to suppliers on an enquiry register or the enquiry abstract (Figure 5.1) could be used. Where a computer is used to maintain a vendor database, records can be kept as data files.

Plant

The contractor's plant requirements will be established in the method statement and programme. They will establish the basic performance requirements of plant and in many cases will have identified specific plant items needed for the works. The duration for which the plant is needed on site will be established from the tender programme. The estimator must firstly compile a 'Schedule of plant requirements', listing the type, performance requirements and durations. This should be separated into:

- Mechanical plant with operator.

- Mechanical plant without operator.

- Non-mechanical plant.

A note must be made on the schedule of additional requirements associated with a particular item of plant, which must be provided by the contractor. A power supply for a tower crane, for example, could be a significant extra cost and temporary access roads for erection purposes may be needed together with foundations.

Further details are necessary for certain non-mechanical plant. A scaffolding schedule must be drawn up by the estimating team in order to provide scaffolding contractors with a clear list of requirements. There are seldom work items in a bill of quantity for temporary works although the preliminaries may give specific requirements such as temporary roof structures or bridging scaffolding to span low-level obstructions. A typical scaffolding schedule is shown in Figure 6.4 (see page 76).

The contractor must consider the intended method of working and programme requirements in the specification for plant. Turnaround of equipment and striking time will dictate the amount of formwork and support and access equipment needed. A balance must be drawn between speed of operation and economy in establishing plant needs and all must be clearly reflected in the plant enquiry.

A list of plant suppliers must be established from companies who can meet the project's requirements. The options available for obtaining plant include:

- Purchasing plant for the contract (in accordance with company policy).

- Hiring existing company owned plant.

- Hiring plant from external sources.

Purchasing plant for the contract

The decision to purchase plant for a particular contract is taken by senior management. Such a decision requires a knowledge of plant engineering and will be made in accordance with the accounting policy of the company. Purchasing of plant is outside the scope of this Code but, for guidance purposes only, the following general factors must be considered when plant is to be purchased for a project and sold on completion:

- Purchase price less expected resale value after allowing for disposal costs.

- Return required on capital invested.

- Cost of finance.

- Cost of maintaining the plant and associated overheads.

- Stock levels of spares.

- The company's policy on depreciation.

- Likely working life of the plant.

- Cost of insurances and taxes, e.g. road fund tax.

- Any tax or depreciation allowances that are available.

- Availability and cost of plant outside the company.

- Accessibility of the site in relation to company depot and servicing centres.

The manner in which the cost of such plant is charged subsequently to a contract will depend upon the accounting policy of the company.

Hiring existing company owned plant

When plant is already owned by the company the estimating department will be provided with hire rates at which plant will be charged to the site. The following list should be regarded, as guidance only, of the items which must be considered in building up hire rates for company owned plant:

- A capital sum based upon the purchase price and expected economic life (this will vary according to the company's accounting policy).

- An assessment of the costs of finance.

- The return required on capital invested.

- Grants and financial assistance available when purchasing the plant.

- Administration and depot costs.

- Costs of insurances and road fund licences.

- Maintenance time and costs and also cost of stocks needed for maintenance purposes.

Hiring from external sources

Where company owned plant is not available, enquiries must be sent to external suppliers for the plant required.

Enquiries for items of plant must either be sent specifying particular machines and equipment that are needed or specifying the performance required from the item of plant. For example, 'tracked excavator with backacter, required to excavate trenches to a maximum depth of 3.00 m, width of excavation 1.00 m, 360° slewing required'.

Enquiries must state:

- Title and location of the work, and address of the site.

- Specification of the plant or work to be done.

- Anticipated periods of hire with start date on site and duration required.

- Means of access, highlighting any restraints or limitations.

- Any traffic restrictions affecting delivery times.

- Anticipated working hours of the site.

- Date the quotation is required.

- Period the quotation is to remain open.

- Whether fluctuating or firm price required, the basis for recovery of fluctuations and the base date when formulae are used for the recovery of fluctuations.

- Discounts offered.

- The person in the contractor's organisation to be contacted regarding queries.

In addition to the basic hire charge per hour or week, the enquiry must seek to establish:

- Cost of delivering and subsequently removal of plant from the site on completion of hire.

- Cost of any operator, over and above the basic hire charge if provided by the hiring company. (If provided by the contractor, the estimator must produce a built-up rate for the operator's costs.)

- Whether the hire rates quoted include for servicing costs. If not, the costs and timing of servicing must be established.

- Any minimum hire periods applicable to the plant and the extent of any guaranteed time.

- Cost of standing time and insurance costs if the plant is retained on site not working for any reason.

The contractor has a responsibility to ensure that suppliers:

- Make every effort to meet the specified 'date required'. If this is not possible the contractor should ensure that he is informed promptly, so that additional enquiries can be sent out in order to maintain a full enquiry list.

- Clear queries as they arise.

- Submit the quotation on time with a clear statement where prices are 'to follow'.

The contractor must record details of all enquiries to plant suppliers on the resources abstract or a plant quotation register, to await the receipt of suppliers' quotations.

Domestic sub-contractors

Sub-contractors will require the same details and information about the sub-contract works as the main contractor requires for his tender. Selection of sub-contractors will take into account skill, performance, integrity, responsibility and proven competence in health and safety matters and for work of similar size and character to the project under consideration and interest in tendering for the particular project. The principles of selective tendering relating to the main contract must be reflected in the number of sub-contract enquiries. The contractor's enquiry to sub-contractors must state:

- Site, and location of the works.

- Name of employer.

- Names of the consultants.

- Relevant contract and sub-contract details.

- Any amendments to the main contract, appendix or sub-contract conditions.

- Whether a fixed or fluctuating contract stating relevant details and rules.

- Daywork rates as required.

- Date quotations must be returned.

- General description of the works.

- Particulars of access, available site plant, site industrial relations policy, storage facilities, etc.

- Where further details and drawings may be inspected.

- Contract period, programme, phasing requirements and method statement details, start date and duration for the sub-contract works.

- Any discounts required.

- A copy of the relevant extracts from the preliminaries and bills of quantity.

- Drawings, schedules and reports where applicable.

- Services or attendance to be provided by the main contractor (if any).

Contractors use a standard format for producing sub-contract enquiries comprising a letter addressed to each sub-contractor, a list of accompanying documents and an abstract of contract conditions adapted for the project. The letter given in Figure 5.3 is a suitable example using the report feature of an enquiry database.

Opportunity must be provided for sub-contractors to see all the relevant drawings and details and every assistance given to ensure that the quotation received is the sub-contractor's best price and not inflated because of inadequate project information.

A simple procedure to avoid errors or misunderstanding is to provide photocopies of all relevant information as far as possible and to delete any irrelevant items not applicable to the enquiry in question. Drawings could be reduced to, say, A3 size to save copying costs and postage; but, if specialists are to take off their own quantities, full-size drawings must be provided. The objective of the enquiry is to provide all relevant information and produce technically accurate quotations.

The contractor has a responsibility to ensure that sub-contractors:

- Make every effort to meet the specified 'date required'. If this is not possible the contractor should ensure that he is informed promptly so that additional enquiries can be sent out in order to maintain a full enquiry list.

Figure 5.3 Typical enquiry letter to sub-contractors

[Date]

To [sub-contractor]

Dear Sirs,

[PROJECT TITLE]
[SPECIALIST TRADE PACKAGE]

We invite you to tender for the above work and enclose the following documents which will form the basis of your offer:

Site plan drawing number:	A/100
Preliminaries pages:	1/1-12
Specification pages:	2/23-32
Bill pages:	4/3-5, 7, 11-13
Drawings numbered:	A/201-205, 304, 305
Daywork schedule:	7/1-3
Health and safety plan:	
Form of tender:	

Your form of tender, priced bills of quantity and daywork rates must be delivered to this address to arrive no later than 12 noon on Wednesday 12 April...

The form of sub-contract will be DOM/1 incorporating all relevant published amendments, and the following:

Payments:	monthly
Discount to main contractor:	2.5%
Fluctuations:	firm price
Liquidated damages:	£950.00 per week
Retention:	5%
Method of measurement:	SMM7
Defects liability period:	six months

We will provide all sub-contractors with water, lighting and electricity services near the work and common welfare facilities on site. Sub-contractors will be required to provide the following services and facilities:

1. Unloading, storing and taking materials to working areas.
2. Power and fuel charges to temporary site accommodation.
3. Clearing-up, removing and depositing in designated collection points on site all rubbish and surplus packing materials.
4. Temporary accommodation and telephones.
5. Day-to-day setting out from main contractor's base lines.

If you have any queries or wish to arrange a site visit please contact the estimator for the project, ...

Would you please acknowledge receipt of this enquiry and confirm that you will submit a tender in accordance with these instructions.

Yours faithfully,

- Clear queries as they arise, in order to avoid quotations marked 'more information required'.

- Submit the quotation on time with a clear statement where prices are 'to follow'.

Labour-only sub-contractors

The contractor may seek quotations from labour-only sub-contractors for certain elements of the work. The decision to use labour-only sub-contractors will take into account company policy, availability of labour in the area, any special requirements concerning site labour policy and market conditions. In this case all items noted under 'Domestic sub-contractors' apply. The contractor must gain further information concerning the status of the labour-only sub-contractor, particularly in regard to whether:

- The sub-contractor holds a current relevant sub-contractor's tax exemption certificate.

- The sub-contractor's insurances are adequate.

In addition the estimator must allow the cost of:

- Any attendances to be provided by the contractor.

- Responsibility for unloading, storage and distribution of materials; waste allowances.

- Supervision to be provided by the contractor.

- Frequency of payments required by the sub-contractor.

- Retention sums relating to the sub-contracted work.

- Any additional costs that may arise once the sub-contractor has completed the works and left the site.

- Any additional risk.

The contractor must record all enquiries to sub-contractors on the resource abstract form (Figure 5.2) or in a domestic sub-contractors' register (see Figure 5.6), to await receipt of the sub-contractors' quotations.

5.3   QUOTATION RECEIPT AND ANALYSIS

The receipt of all quotations must be recorded methodically on the appropriate quotations register. Upon receipt of all quotations a detailed examination is necessary to verify that they have been made in accordance with the requirements of the enquiry. Even though precise information has been given to suppliers and sub-contractors, the contractor will frequently find that alternative offers have been submitted. In all cases the contractor must ensure that all additions, extensions and collections are correct and that no pricing errors have occurred.

Materials

Responses from suppliers are recorded on the resource abstract sheet or in a materials quotations register. Rates are entered on a 'Materials comparison form' (see Figure 5.4) for use during the pricing stage. Comparison forms are being replaced by computer systems which accept lists of rates. The computer method allows direct links with the resources used to price the work. Discrepancies and divergence from the enquiry must also be recorded on the register for later evaluation, before selection of the quotation to be used in pricing unit rates. Quotations must be checked to ensure that:

- The materials comply with the specification.

- The materials will be available to meet the requirements of the construction programme.

- No special delivery conditions have been imposed by the supplier.

- The method of delivery complies with the contractor's requirements and intended method of unloading and handling on site.

- The conditions contained in the quotations do not amount to an alternative offer being at variance with the terms and conditions of the enquiry.

- The quotation is valid for the required period.

- Prices are given for small quantities and other items if required.

- Discounts comply with the requirements of the enquiry.

- Requirements concerning fixed or fluctuating prices are complied with.

KEYSTONE CONSTRUCTION LTD

MATERIALS COMPARISON FORM

Project Helix Laboratories, Westfield

Ref.	Category	Type	Approx. quantity	Unit	Supplier 1	Supplier 2	Supplier 3	Supplier 4	Supplier 5	Rates used	Remarks
					Target	Wilson	Dixon				
M01	Aggregates	Hardcore	12,000	t	7.00	7.15	7.00			7.00	
M02		Type 1 sub-base	16,000	t	6.75	9.05	8.90			6.75	
		comparative total			192,000	230,600	226,400				
					MBC	Wilson	Montel				
M03	Concrete	20 N	250	m³	45.25	48.50	47.25			45.25	cement replacement
M04		30 N	300	m³	52.60	55.20	55.35			52.60	cement replacement
		comparative total			27,092	28,685	28,417				
					Dexter	Opus					
M07	Masonry	Facing bricks type A	55	th	335.00					335.00	
M08		Blocks 140 mm	6500	m²	9.65					9.65	

Figure 5.4 Materials comparison form

The quotation must be examined to ensure completeness to identify any alternative offers of material which may not be acceptable or where items have been priced 'as price list'. Standard price lists are not always simple to apply to a particular project, and care should be taken to arrive at the appropriate price when using such lists. A number of discounts are usually available and must be taken by the estimator either during the pricing stage or on the summary form for materials. Suppliers also have additional charges which are not always evident in price lists.

Plant

Responses from plant suppliers are recorded on the resource abstract sheet or in a plant quotations register. Rates are entered on a 'Plant quotations register' (see Figure 5.5) for use during the pricing stage. Comparison forms are being replaced by computer systems which accept lists of rates. The computer method allows direct links with the resources used to price the work. Discrepancies and divergence from the enquiry must be identified and also recorded on the register. It must be borne in mind that where a performance specification is provided for the plant, quotations must be carefully checked to ensure that the plant offered meets the requirement defined at the enquiry stage.

Quotations must be scheduled to ensure that:

- The plant complies with the project requirements.

- Plant is available to meet the needs of the construction programme.

- Delivery and collection charges can be identified.

- Where applicable, all operator costs are included and the operators will comply with the intended working hours of the site.

- Any attendance or supplies to be provided by the contractor are clearly identified.

- Maintenance responsibilities and charges are identified; tyre wear is a particular case to check.

- Maintenance liabilities are identified.

- The quotation is valid for the required period.

KEYSTONE CONSTRUCTION LTD		PLANT QUOTATIONS REGISTER							Project	Helix Laboratories, Westfield

Ref.	Plant item	Unit	Basic hire rate	Operator cost	Fuels	Maintenance costs	Delivery and collection	Totals	Rates used	Remarks
	PLANT DEPARTMENT									
	JCB3CX	hr	5.00	8.00	1.50	1.50		16.00		
	20T backacter	hr	7.50	8.00	3.00	2.00		20.50	20.50	
	Forklift	week	180.00	inc	30.00	15.00		225.00	225.00	driver in proj o/heads
	DOLPHIN PLANT									
	JCB3CX	hr	11.50	inc	1.50	1.50		14.50	14.50	fuel not quoted
	20T backacter	hr	16.50	inc	3.00	2.00		21.50		fuel not quoted
	Forklift	week								not available

Figure 5.5 Plant quotations register

- The quotation conforms to the terms and conditions of the enquiry and does not represent an alternative offer.

- Requirements concerning fixed or fluctuating prices are complied with.

Domestic sub-contractors (including labour-only sub-contractors)

Responses from sub-contractors must be recorded on the resource abstract sheet (Figure 5.2) and domestic sub-contractors register (shown in Figure 5.6), the latter being used as a sub-contract comparison form. As for other resources, computer systems have eliminated the need for manual comparison reports. Any divergence or discrepancies from the enquiry must be identified and highlighted for later evaluation at the time of selection of the quotation to be used in pricing unit rates. Quotations should be checked to ensure that:

- Work described in the quotation complies with the specification and bills of quantities.

- All items have been priced.

- All items are priced correctly and in accordance with the unit of measurement billed.

- Unit rates are consistent throughout the quotation.

- Rates are realistic and comparable with those of competitors.

- The quotation is not an alternative offer and that the sub-contractor has accepted the terms and conditions of the enquiry.

- Any attendance required from the contractor is clearly defined and acceptable.

- Any ambiguities regarding responsibility for unloading, storage, protection or clearing up are resolved.

- Discounts comply with the requirement of the enquiry.

- All supporting information, such as schedules of rates and programming information, are provided.

- Requirements covering fixed or fluctuating prices are complied with.

DOMESTIC SUB-CONTRACTORS' REGISTER

KEYSTONE CONSTRUCTION LTD

Project: Helix Laboratories, Westfield

Trade: CFA Piling

Ref.	Name and telephone number	Date of quotation	Quotation amount £	Gross checked BQ total £	Discount %	Net BQ total £	Firm price allowance %	Amount £	Remarks
1	BBS 01237 232456	11 Apr	36,227	40,079	2.50	39,077		Fixed	One pile test missing now added to quotation
2	Dudley 01321 765433	10 Apr	42,200	42,200	2.50	41,145		Fixed	
3	Ross 01761 986782	11 Apr	incomplete						Unable to offer piles to specification
4	Robinson 01324 678912								Price not yet available
5									
6									
7									
8									

Figure 5.6 Domestic sub-contractors' register

6. PLANNING AND TEMPORARY WORKS

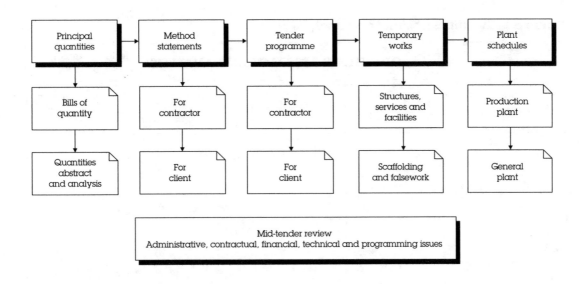

6.1 PRINCIPAL QUANTITIES

The first part of the planning stage is to extract the principal quantities from the bills of quantity for the main elements of work. An experienced estimator or planning engineer is aware of the operations likely to form important and critical items in a programme. The amount of formwork, for example, is more relevant to a programme than the volume of concrete to be placed. This is because concrete can be placed quickly with large gangs or with the assistance of a concrete pump, whereas formwork outputs are limited by the number of skilled carpenters employed on the site and the need to re-use materials wherever possible.

Principal quantities are usually provided in bills of quantity for civil engineering work but rarely for building. It is not too difficult to abstract quantities from any bills of quantity providing ancillary items and labours are ignored. Computers on their own are unable to assist in this appraisal because estimating software is unable to differentiate between significant and insignificant items. In order to overcome this hurdle, contractors may add an activity or operation code to each item entered in estimate file.

Figure 6.1 shows a typical list of principal quantities together with outputs and gang sizes.

6.2 METHOD STATEMENTS

It is essential that an early meeting is held between the estimator and those responsible for the programming and construction of the

Figure 6.1 Quantities abstract and analysis

KEYSTONE CONSTRUCTION LTD	QUANTITIES ABSTRACT AND ANALYSIS						Tender number	
Ref.	Description	Quantity	Unit	Output hrs/unit	Total time		Gang size	Duration wks
					hrs	man wks		
3/3	Formwork to foundations	1126	m^2	1.75	1971	43.79	4	11
3/5	Bar reinforcement to founds	13.5	t	25	338	7.50	2	4
3/1	Concrete in foundations	200	m^3	1.65	330	7.33	4	2
3/8	Blockwork below dpc	255	m^2	0.75	191	4.25	2	2
3/14	Structural steelwork	55	t	15	825	18.33	4	5

project, to set out initial thoughts on the method of construction. In a smaller organisation, all of these roles may be undertaken solely by the estimator.

Method statements are written descriptions of how operations will be carried out and managed. They should not only deal with the use of labour and plant in terms of types, gang sizes and expected outputs, but also include the way in which the project will be organised. It is clear that a contractor will gain a competitive advantage by finding the most cost-effective or innovative systems for temporary works and materials handling. This appraisal of the scope of works and method could involve changes to the design which may be offered as alternative tenders.

The primary purposes of the method statement are:

- To establish the principles on which the estimate is based.

- To acquaint construction personnel of the resource limits that have been allowed in the estimate and to describe the method of working envisaged at tender stage.

Clients or their consultants often request method statements from contractors at tender stage. This provides the client with confidence in the contractor's ability to find and overcome construction difficulties. Other reasons for seeing a contractor's method statement include: satisfying statutory safety requirements, bringing out interface problems which might exist with other contractors on site, and to demonstrate compliance with the client's overall objectives such as ensuring production is not interrupted in a factory.

The disadvantage for clients is that a method statement adds another criterion to the factors used to choose a contractor. Good practice for selective tendering recommends selection on price alone.

The following points must be considered when deciding on the method of construction:

- Site location and access.

- Ground conditions.

- Degree of repetition.

- Shape of the building.

- Extent that the building works occupy the site and, therefore, the available space for storage, accommodation, movement on the site and temporary works.

- Adjacent buildings, structures and other contractors on the site.

- Company's experience of the type of building required and the location.

- Availability of labour.

- Availability of materials.

- Extent of specialised work and its relationship to the general construction.

- Work packages to be sub-contracted.

- Amount that design indicates construction method, i.e. restraints, formwork striking times and special sequence of construction.

- Requirements for phasing the works.

- Health and safety issues arising from the pre-tender health and safety plan and the contractor's own appraisal of the works.

- Maintaining standards and compliance with the specification.

- Plant requirements.

Decisions concerning the resources to be used on the project will take into account such factors as:

- Location and availability of labour and management within the company.

- Cost of recruiting additional labour, its availability, quality and quantity.

- Amount of work to be sub-contracted.

- Plant available within the company.

- Availability of plant outside the company.

- Availability of materials, including long delivery items.

- Current and future projects in the area which may affect the supply of basic resources.

- Quality of workmanship required.

- Special requirements of the project, i.e. special plant or skills needed.

- Overlap of operations needed to meet programme requirements.

- Materials handling on site, storage, distribution and waste, time span of the project and seasonal influences on method of construction.

- Quality and complexity of the work.

For successful tenders, the method statement will outline the sequence and methods of construction upon which the estimate is based. It should indicate how the major elements of work are to be dealt with and should highlight areas where new or difficult methods are necessary or intended. It should be supported with details of cost data, gang sizes, plant requirement and supervision requirements.

A pre-tender health and safety plan drawn up by the planning supervisor will be provided to all tenderers. This will incorporate an identification of risks arising from the nature of the work and the design. The design team must attempt to minimise risks to health and safety during the construction stage and when the building is used and demolished.

6.3   TENDER PROGRAMME

The tender programme is a vital document for the contractor. If the contract is secured, this programme represents the contractor's intentions at the time of tender and upon which the pricing of the works was based. Many standard forms of contract require the contractor to submit a copy of his programme at an early stage of the contract and the programme submitted may well be the tender programme. As it reflects the contractor's intentions, based upon the tender information provided, it is appropriate that it should be the document used.

The contractor will not always be required to submit a programme with his tender and may only be required to confirm by signing the form of tender that all the works will be executed in accordance with the commencement and completion dates stated. However, good estimating practice dictates that the contractor must prepare a tender programme to:

- Verify that the date given for completion is possible.

- Maximise the efficiency of resources.

- Price the time-related elements, in particular staff and site accommodation, temporary works and general plant.

- Establish the method and sequence of working.

- Determine cost increases for firm price tenders.

- Identify work affected by seasonal weather changes such as bulk excavation, drying the internal fabric of the building and landscaping.

- Consider making an offer with a shorter contract duration.

The tender programme usually develops in two stages. A preliminary programme is drawn up at an early stage of the project once the project information has been assimilated. This must take into account:

- Contract commencement and completion dates.

- Client requirements for phasing or stage completion.

- Other stated requirements or sequence of work, i.e. delivery of client's equipment.

- Work to be carried out by directly employed labour.

- Work to be carried out by domestic sub-contractors.

- Work to be carried out by nominated sub-contractors.

- Key items to be supplied by nominated suppliers.

- Timing for temporary works items such as scaffolding.

The preliminary programme will thus identify the parameters of the project and the main resources needed. It will establish key dates for major portions of the work and provide basic information which can be used when obtaining quotations for materials, plant and work to be sub-contracted.

A clear basis now exists on which construction method and sequence can be developed, following the clarification of any queries by the consultants and the site visit. When such queries and constraints have been resolved, the tender programme can be produced from the preliminary programme, taking into account the agreed construction method and sequence. The method statement can also be finalised.

The completed tender programme must clearly identify the contractor's intentions concerning the construction of the work. It must be used to check and balance the resources needed on the project, against those already contained in the estimate. It is an important document to be considered at the final review stage.

When design is incomplete at the tender stage, the contractor will find it beneficial to incorporate in the tender programme a series of key dates by which design information (or instructions regarding PC and provisional sums) must be given. It may well be that the contractor's estimate is based on early ordering of materials to obtain price advantage and/or long delivery periods of materials. It is essential that design information is made available to reflect these intentions and secure the prices contained in the contractor's estimate.

A project overheads programme is superimposed on the tender programme in order to quantify the time-related resources. Figure 10.2 (see page 134) shows durations for site staff, accommodation and general plant.

When a client or his consultants require the submission of a programme with a tender, the contractor produces a simplified version of his tender programme with four aims:

- To make a well-presented submission.

- To allow opportunities for changes to be made by the site team when the construction programme is produced.

- To offer an overall duration which meets the client's general requirements and in cases where time is critical, offer early completion; either because it is feasible or to gain competitive advantage. (With non-adversarial forms of contracting such as partnering, where co-operation is encouraged, and price is not an overriding criterion, a contractor can look for alternative methods, sequences and resources in order to offer an optimum solution.)

- To establish dates by which design information will be required to meet the programmed completion date.

The programming technique used will reflect the complexity of the contract, but comment on programming procedures and techniques is outside the scope of this Code. It must be recognised that contractors might attempt to disguise any ideas which could be adopted by competitors.

6.4 TEMPORARY WORKS

The temporary works element of a project is designed by the contractor and as such is an important opportunity for improving the competitiveness of his tender. Innovative solutions for supporting structures or access equipment can make the difference between winning or losing the contract. As an example, it is worth considering mobile access equipment which can be cheaper to use than fixed scaffolding. There is a question of scale, and the advantages will depend on the size and complexity of the scheme.

The estimating team will mark up a site layout drawing to show the position of access routes, restrictions, areas for storage and accommodation, temporary service connections and distribution, temporary spoil heaps, cranage, batching plant, hoardings and scaffolding. Schedules are commonly produced for traffic control and scaffolding, for which prices are obtained from specialist contractors. Figure 6.2 illustrates a scaffold schedule which provides a clear basis for obtaining quotations at tender stage.

The costs for temporary works are calculated and summarised in Chapter 10 on the project overheads schedule forms 'Temporary works' and 'Mechanical plant' (see Figures 10.9 and 10.10 on pages 141 and 142).

Figure 6.2 Scaffold schedule

KEYSTONE CONSTRUCTION LTD	**SCAFFOLD SCHEDULE**		Tender number	

Ref.	Type	Description	Quant.	Programme	
				From	To
1	Ext independent scaffold	South wing three elevations		wk 6	wk 26
2		Part north wing around stairtower		wk 6	wk 26
		Note: boarding for access to windows and roof parapet			
		Note: allow for fixed ladder access to both areas			
3	Internal scaffolding	To lift shaft		wk 12	wk 18
4	Temporary roof	South wing only		wk 8	wk 24
		Note: waterproof sheeting down to parapet level			
		Note: allow for temporary rainwater installation			
5	Loading platform	To south wing at parapet level 4.00 × 5.00 m		wk 6	wk 26
6	Hoist towers	Not required			
7	Roof edge protection	Electrical sub-station only		wk 24	wk 29
8	Debris netting	North wing stairtower scaffolding		wk 6	wk 26

Related documents	Drawings	
	Specification	
	Bill pages	

6.5 PLANT SCHEDULES

The plant element of an estimate is classified as production plant and general plant. Production plant is that which is priced in the measured work sections of a bill of quantities and general plant is that part of a contractor's project overhead calculation for plant excluded from unit rates. General plant is available as a general facility on site across a wide range of operations. In Chapter 10, general plant appears on two forms 'Mechanical plant' and 'Non-mechanical plant' (see Figures 10.10 and 10.11 on pages 142 and 143). Production plant is listed on the 'Plant summary' form in Chapter 11 (see Figure 11.2 on page 156).

For some projects, particularly civil engineering works, it is not appropriate to build up unit rates which include plant, because whole operations need to be examined. An excavator may be used, for example, to dig a trench for drains and manholes and assist in laying pipes and manhole rings. In this case the number of hours of work for the excavator is determined by reference to a programme.

If plant has been included in an estimate through unit rate pricing, the plant summary will indicate the total duration for each piece of equipment. This is then reconciled with the tender programme to see if sufficient time has been included in the estimate to allow for continuous periods on site. Typically a contractor needs to add the cost of standing time to plant, such as rollers for compacting fill materials – an operation which takes place intermittently.

6.6 MID-TENDER REVIEW

In order to maintain progress and manage the preparation of a tender, the estimator should report regularly to management. Formal meetings are not usually necessary, but for large projects a mid-tender review meeting provides an opportunity for the estimator to explain his approach and others to give their ideas and expectations. The following simple agenda is used not only to look at what has been achieved but to make the best use of the remaining time and introduce any new impetus.

1.0 Administration
- staff allocation and responsibilities
- site visit
- enquires to suppliers and sub-contractors
- tender amendments
- presentation documents
- delivery of tender

2.0 Contractual
- check on terms and amendments
- check sub-contract arrangements

3.0 Financial
- insurances
- bonds
- cash flow

4.0 Technical
- design appraisal
- temporary works
- plant
- special risks
- safety issues
- alternatives

5.0 Programme
- type
- alternative methods and sequence
- input from specialists

6.0 Queries
- decide which queries to raise with client
- consider whether more information is required
- consider qualifications to tender
- consider alternative tender
- look at need for extension to tender period

7. ALL-IN RATES AND UNIT RATES PRICING

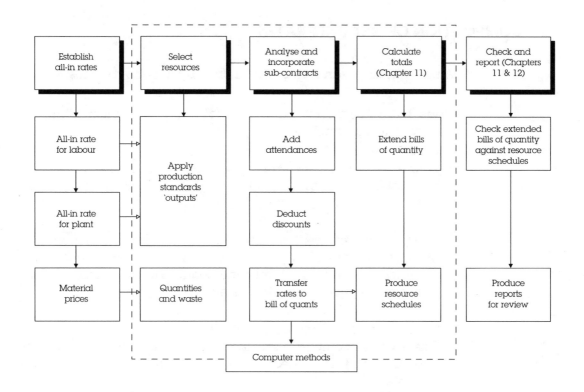

7.1 PRICING STRATEGY

In Chapter 6 the project was analysed in order to list the principal items of work and examine the methods to be adopted. The estimator now needs a clear strategy for pricing the bills of quantity. It is unlikely that items will be priced in the order they appear in the bills because a better understanding of activities can be gained by pricing one trade at a time. This is becoming increasingly popular with computer estimating systems which readily sort the bill items into trade order (similar items). Computers also allow resources to be entered either through a resource build-up screen for each item or with the aid of a 'spreadsheet' type comparison system where like trade items can be viewed in a single table.

If it is known that quotations for materials will be delayed, the estimator can price labour and plant first, and return to part-priced items later when quotations are available. On the other hand, 'typical' materials prices may be used during the pricing stage. Computers allow late adjustments to be made and all affected items will be changed.

All these techniques are derived from manual pricing methods but are now faster and more accurate with the use of computers.

7.2 COMPOSITION OF NET UNIT RATES

A net unit rate for an item of work is built up in three distinct stages:

Stage 1 The establishment of all-in rates for the key items that will be incorporated. It includes:

- A rate per hour for the employment of *labour*. Different rates will be established for the different categories of labour that will be used on a project.

- An operating rate per hour (or per day, per week, etc.) for an item of *plant*. This is for plant supplied with or without an operator and the rates are established from the contractor's own data or from quotations received from plant hire organisations.

- A cost per unit of *material* delivered and unloaded at the site. This involves comparison of the various quotations received for materials and the selection of one of these for use in the estimate.

The cost of labour and some plant items are established after the visits to the site and consultants. It is also necessary to complete the tender programme and method statement before finalising these prices. It will be necessary to await receipt of quotations in respect of materials, some plant items and sub-contracted work. With computer-aided estimating systems the estimator can use pricing with 'notional' rates because they can be adjusted at any stage in the preparation of an estimate.

Stage 2 The selection of methods and *production standards* from the contractor's data bank or other sources. These standards are then used, in conjunction with the all-in rates calculated in Stage 1, to calculate net unit rates which are set against the items in the bills of quantity. Alternatively, rates received from sub-contractor's are used.

Stage 3 The incorporation of rates from specialist trade contractors, including those offering 'labour-only' services, either producing the whole or part of a rate.

[*The calculation and addition of project overheads is a separate and subsequent operation dealt with in Chapter 10 and the preparation of reports for consideration by management are considered in Chapter 12.*]

7.3   ESTIMATING ALL-IN RATES FOR LABOUR

Items for consideration in all-in rates for labour

Labour costs arise in two areas. Costs associated with the Working Rule Agreement (WRA) and certain overhead costs incurred by the employer. Other costs, which will be variable and are specific to a project, or some time-related costs, will be contained in the project overheads.

Labour costs normally contained in the all-in rate

- Guaranteed minimum wages. — Basic rates in NWR 2 for each class of operative.

- Contractor's bonus allowance. — Bonus used to retain operatives which is not self-financing.

- Inclement weather allowance. — Normally included by paying full weekly wages.

- Non-productive overtime costs. — NWR 7 gives the rules for calculating overtime rates.

- Sick pay allowance. — NWR 16 gives amount and qualification rules.

- Trade supervision. — Proportion of non-productive time by supervisors often considered in project overheads.

- Working Rule Agreement allowances. — NWR 3 and 4 list extra payments for labourers engaged in particular activities.

- CITB training contributions. — Normally levied at 0.25 per cent of entire payroll, and 2 per cent for self-employed and labour-only sub-contractors.

- National Insurance contributions. — Employer contributions are percentage of weekly earnings depending on which of six bands apply.

- Holiday credits. — Annual and public holidays.
- Tool allowances. — Given in NWR 18.
- Severance payments. — Statutory scheme.
- Employer's Liability insurance. — Often considered in project overheads.

Labour costs normally contained in project overheads

• Daily travel allowance.	NWR 14 provides a scale of allowances for daily travel one way. Some employers provide transport which reduces the allowances.
• Periodic leave and lodging allowance.	Travelling and lodging allowances are given in NWR 15.
• Supervision.	Usually priced using tender programme.
• Attraction money.	Needed in remote areas or sites which might experience shortages of local labour.

General

It may not be possible to determine all of these factors with accuracy at an early stage of estimating, particularly as the volume of labour is not quantified. There are benefits in separating them from the calculation of all-in rates and for necessary allowances to be made in the later stages of the estimating process, in the project overheads.

It is a matter of opinion and company preference where many of these items are priced. The distinction drawn between items set out in all-in rates and project overheads should not be regarded as mandatory. The important consideration is that due recognition must be paid to all items to establish the true costs involved and that adequate allowances are made in the estimate.

All-in rates are built up on a weekly or annual basis, or the time period relating to a particular contract. In this Code the all-in rate is calculated on an annual basis.

7.4 CALCULATING AN ALL-IN HOURLY RATE FOR LABOUR

There are three stages in the calculation:

1. Determine the number of working hours that an operative is expected to work during the one-year period.

2. Calculate the cost per year for wages and the cost of each item used from the list contained in Section 7.2.

3. Summarise the individual costs obtained in (2) and calculate the all-in rate per hour by dividing the total costs in (2) by the number of hours in (1).

Alternatively, for a very large project, it may be desirable to make a special calculation based on the anticipated construction period.

- It is emphasised that the calculations in the following are examples only.

- Details vary according to:

  — the actual trade of the operative (whether craft operative, labourer or mechanical plant operator)

  — the firm

  — the area

  — the industrial and legal conditions in force at any time

- In this example the calculation is based on a one-year employment period, and could apply to all projects for which tenders are to be submitted. It is for a craft operative.

- Amendments must be made each time there is a variation in the cost of one of the factors included in the calculation, or when further factors are introduced. Alterations must be made as soon as variations are promulgated, though there may be a period of time before they come into effect.

- The whole calculation should be revised regularly.

1 Determination of hours worked

The number of hours worked during the calendar year, i.e. January to December, will depend upon the hours worked per week during the summer and winter periods, with adjustments for annual holidays and public holidays. The hours worked will vary between different companies and some variation in hours can also be expected between firms operating in the north and south of the country, due to the available amount of natural daylight hours in the winter period. Local customs and availability of labour also affect the number of hours worked. A company may agree to work hours suitable for a particular type of contract, due to special requirements of the employer or by special agreement with its employees.

The Working Rule Agreement, states that a 39-hour week should be worked throughout the year as follows:

- Mondays to Thursday – 8 hours per day.

- Fridays – 7 hours.

For calculation purposes only, the hours used are typical of those worked on many sites. Although the 'summer period' defined in the Working Rule Agreement is 1 April to 30 September it is common to assume that a longer working day can be achieved during British Summer Time, assumed to be 30 weeks long.

An operative's annual and public holidays have been taken as listed in Table 7.1 and the total number of days are as agreed in the Working Rule Agreement. In some regions the actual day to which the holiday is allocated will vary according to local tradition.

The hours worked by an operative during the summer and winter periods are detailed in Table 7.2.

Table 7.1 Annual and public holidays

	Annual holidays (21 days)	Public Holidays (8 days)
Winter	7 working days at Christmas	Christmas day Boxing day New Years Day
Summer	4 working days at Easter 10 working days (2 calendar weeks)	Good Friday Easter Monday First Monday in May Spring Late summer

Table 7.2 Calculation of hours worked

Summer period			Winter period		
Starting time		8.00 am	Starting time		8.00 am
Lunch period		1.00–1.30 pm	Lunch period		1.00–1.30 pm
Finishing time	Mon–Thurs	5.30 pm	Finishing time	Mon–Thurs	4.30 pm
	Friday	4.30 pm		Friday	3.30 pm
		Hours			*Hours*
Working week		44	Working week		39
For 30 weeks		1320	For 22 weeks		858
Deduct			*Deduct*		
10 days summer holiday		(88)	7 days Christmas holiday		(55)
4 days Easter holiday		(35)	3 days public holiday		(23)
5 days public holiday		(44)			
Total hours for summer		**1153**	*Total hours for winter*		**780**

Gross hours available for work	$1153 + 780 = 1933$
Deduct allowance for sickness, say 8 days in winter	(62)
Net hours available for work (basic hours)	1871

Inclement weather

The time lost for inclement weather will vary according to the type of work, season of the year and geographical area. An average allowance is used and any necessary adjustment is made for exceptional situations in the project overheads. Say the time lost due to inclement weather is 2 per cent, i.e. approximately 37 hours.

$$\text{Actual hours worked} = \text{Basic } less \text{ inclement weather}$$
$$= 1871 - 37 = 1834$$

2 Calculation of annual employment costs

Guaranteed minimum wages and emoluments (Annual Costs)

Note: All figures quoted are those *current* as at 27 June 1996.

(a) *Guaranteed minimum wages*

Hours worked per year (basic hours) = 1871

The weekly rate of pay at 27 June 1996 gives guaranteed minimum weekly earnings (basic rate) of £178.62 per 39 hour week.

$$\text{basic annual wages} = 1871 \text{ hrs} \times £4.58/\text{hr}$$
$$= £8569.18$$

(b) *Contractor's bonus allowance*
At, say, £20 per 39 hour week.

$$1871 \text{ hrs} \times £0.51/\text{hr} = £954.21$$

(c) *Inclement weather allowance*
This is included in guaranteed minimum wage above.

(d) *Non-productive overtime costs*
Overtime rates are paid for time worked in excess of the normal working hours. This allowance is for overtime worked as normal practice; the cost of any additional special overtime will be dealt with separately in the project overheads.

In our example calculation, the total hours worked by an operative during the summer and winter periods are shown in Table 7.3.

Working weeks in summer:
30 *less* 4 weeks for annual and public holidays = 26
Non-productive overtime for summer period = 26 × 2.5 hrs
= 65 hrs

Table 7.3 Total hours worked during summer and winter periods

	Mon	Tue	Wed	Thu	Fri	Sat	Sun	**Total**
Summer period								
Hours worked	9	9	9	9	8	–	–	44
Basic hours	8	8	8	8	7	–	–	39
Overtime hours	1	1	1	1	1	–	–	5
Non-productive hours	0.5	0.5	0.5	0.5	0.5	–	–	2.5
Winter period								
Hours worked	8	8	8	8	7	–	–	39
Basic hours	8	8	8	8	7	–	–	39
Overtime hours	0	0	0	0	0	–	–	0
Non-productive hours	0	0	0	0	0	–	–	0

The overtime allowance is calculated as follows:

Total for whole year = 65 hrs

Therefore, cost of non-productive overtime at a basic hourly rate, per 39 hour week, of £178.62/39 = £4.58 is:

£4.58 × 65 = £297.70

(e) *Sick pay*
For 8 days of sickness the first 3 qualifying days are not due for payment. The assumed number of payable days lost due to sickness is 5. The statutory rate of payment is £12.10 per day.

Cost of sick pay = 5 × £12.10
= £60.50

If a private insurance scheme is used instead of a statutory scheme, the cost of premiums should be included here instead of the statutory allowances.

(f) *Trade supervision*
The number of trades foremen to operatives will vary from company to company and in accordance with the needs of a project. Assume that:

- There is one trades foreman for every eight tradesmen.

- Half his time is spent working and half on supervisory duties.

- His rate of pay is £0.32 per hour above the trade rate.

Hourly cost for the gang is

$$
\begin{array}{lll}
\text{1 trades foreman} & £4.58 + £0.32 = & £4.90 \\
\text{8 tradesmen} & £4.58 = & £36.64 \\
& \textit{Total} & £41.54
\end{array}
$$

Allowing for supervision time, the effective hourly rate for working operative is:

$$£41.54 \div 8.5 = £4.89$$

Therefore, additional hourly cost of supervision is:

$$£4.89 - £4.58 = £0.31/hr$$

Basic hours worked per year $= 1871$. Therefore, cost of trade supervision is:

$$1871 \times £0.31 = £580.01$$

It is normal to use a firm's individual arrangements in basic calculations and adjust in the project overheads (if necessary for a particular project).

(g) *Working Rule Agreement allowances*
Operatives exercising special skills or working in particular circumstances are entitled to special allowances under the Working Rule Agreement. Examples include discomfort, inconvenience or risk, continuous extra skills or responsibility, intermittent responsibility, tool allowances, special provisions or servicing of mechanical plant, storage of tools and clothing. The amount allowed is, therefore, variable according to responsibility and skills. The following is used as an example for the driver of a rough terrain forklift truck up to 3000 kg capacity.

The allowance (for Working Rule Agreement allowances) to be paid is £14.46 per week, or £0.37 per hour. Therefore, annual cost of WRA allowance is:

$$1871 \, hrs \times £0.37 = £692.27$$

Since the extra payments form part of an operative's basic rate, they must be included in the overtime calculation, as follows:

$$65\,\text{hrs overtime} \times \pounds0.37 = \pounds24.05$$

The extra payments are mainly supplements for general operatives and not craft operatives.

Overheads in employing labour

(h) *Training*

Many companies used to have their own training departments and the costs were normally carried in head office overheads. Training is still undertaken whether at the company's cost or sponsored by Construction Industry Training Board (CITB) schemes. The operating costs of the CITB are recovered by way of a levy (0.25 per cent) of annual payroll rates applied to managerial, clerical and operative employees.

Where labour-only services are used, the levy is based on a percentage of payments made for labour-only at a rate of 2 per cent. Firms whose payroll together with payments for labour-only services are less than a lower limit during the year are excluded from payment of the levy.

Each company will vary in the amount actually paid to the CITB but for directly employed labour the 0.25 per cent payment would apply.

(i) *National Insurance contributions (not contracted out)*

There are different levels of contributions depending on the employee's gross weekly earnings. From 6 April 1996, for weekly earnings in the range £110.00–£154.99 the Employer's Contribution is 5 per cent on whole earnings. Since most operatives' earnings will fall into the next range £155.00–£209.99, the higher rate of 7 per cent applies. (It should be noted that National Insurance contributions are payable on all payments, including any productivity bonus which may not be included in the all-in labour rate. It may therefore be necessary to make an assessment of the anticipated value of productivity bonus payments and apply the appropriate percentage National Insurance to the total earnings value.)

(j) *Holiday Credits (including Death Benefit Scheme)*

Employers make weekly payments on behalf of employees for annual holidays and the cost is the same for tradesmen and general operatives.

The cost of an annual holiday stamp is £19.60 per week (including a £2.05 contribution towards the Death Benefit Scheme). However, stamps do not have to be paid during holiday weeks, therefore the cost per year is

$$47 \times £19.60 = £921.20$$

The cost of public holidays must be met by the employer. Paid public holidays must be allowed by the estimator based on the guaranteed minimum wage for each grade of employee for the eight public holidays each year. The cost per year, therefore, is:

$$£198.62 \div 5 \times 8\,(days) = £317.79$$

(k) *Tool allowances*
Although tool allowances are not part of an operative's basic wage, they are subject to National Insurance contributions and the deduction of tax. The tool allowances given in WRA 18 are not taken into account in the calculation of overtime. A carpenter, for example, is entitled to £1.94 per week for the provision, maintenance and upkeep of tools which he/she provides. Most other trades receive £0.99 per week.

(l) *Severance pay and sundry costs*
For the following indeterminate factors allow 2 per cent on labour costs:

• Severance pay.

• Loss of production during notice.

• Absenteeism (the cost of National Insurance, pensions and holidays with pay being spread over a smaller number of working hours than the normal conditions assumed).

• Abortive insurances (paying stamps for operatives who work on Monday in order to have cards stamped but who are absent subsequently).

The percentage allowed will vary from firm to firm according to experience.

(m) *Employer's Liability and Third Party Insurance*
For Employers Liability and Third Party Insurance, allow 2 per cent on labour. This percentage will vary according to the firm, the insurance company, the company's insurance record, and type of work and size of the contract. It is becoming increasingly common for companies to express all their insurance costs as a percentage of

turnover (or in the case of any particular tender, tender price) and it may be appropriate to include this calculation in the project overheads schedule when the full value of the project is known.

3 Summary

A summary of the total annual cost rate for a craft operative for the period 1996–97 is presented in Table 7.4

$$\begin{aligned}
\text{All-in hourly rate} &= \text{Annual cost per craft operative} \\
&\quad \div \text{Number of productive hours worked} \\
&= £13,061 \div 1834 \\
&= \mathbf{£7.12}
\end{aligned}$$

Figure 7.1 shows a typical all-in rate calculation produced using spreadsheet software. Estimators use this approach for rapid changes to certain variables such as bonuses and the number of hours worked each week. There are similar facilities in estimating packages which link this calculation to labour resources and remain linked to the build-up of an estimate.

Table 7.4 Annual costs per craft operative, 1996–97

Guaranteed minimum wages		8569
Contractor's bonus allowance		954
Inclement weather allowance		*Included*
Non-productive overtime costs		298
Sick pay allowance		60
Trade supervision		580
Working Rule Agreement allowances (generally for labourers not craft operatives)		*Not applicable*
	Sub-total	**£10,461**
CITB training contributions (0.25% on annual payroll £10,461)		26
National Insurance contributions (7% of average earnings £200 × 52 wks)		728
Holiday credits		1239
Tool allowances		100
Sub-total		12,554
Severance payments (2% of labour costs)		251
	Sub-total	**£12,805**
Employer's Liability insurance (2% of labour costs)		256
Annual cost per craft operative	*Total*	**£13,061**

7.5 CALCULATION OF NET UNIT RATES

Principles

In calculating unit rates for inclusion in the bills of quantity careful consideration must be given to every factor which may influence the cost of the work. Some of the common factors are given in the following text.

Figure 7.1 Calculation of all-in rate for labour

| KEYSTONE CONSTRUCTION LTD | CALCULATION OF ALL-IN RATES FOR LABOUR | Tender number |
| | | Date |

Description		Data entry		Craftsman	Labourer
Summer period	Number of weeks	30			
	Weekly hours	44			
	Total hours		1320		
	Days annual hols	14			
	Days public hols	5			
	Total hours for hols		−167		
Winter period	Number of weeks	22			
	Weekly hours	39			
	Total hours		858		
	Days annual hols	7			
	Days public hols	3			
	Total hours for hols		−78		
Sickness	Number of days (say winter)	8	−62		
Total hours for payment			**1870**		
Allowance for bad weather (%)		2	37		
Total productive hours			**1833**		
		Craftsman	Labourer		
Annual earnings	Guar Min Wage	£178.62	£147.81		
	Attraction bonus	£20.00	–		
	Total weekly rate	£198.62	£147.81		
	Hourly rate of pay (39th)	£5.09	£3.79		
	Annual earnings = 1,870.40 × hourly rate			9520.34	7088.82
Additional costs	Non-productive overtime (time + half only)				
	Hours per week (summer)		2.50		
	Hours per week (winter)		0.00		
	Hours per year (summer)		65.50		
	Hours per year (winter)		0.00		
	Cost of non-prod overtime			299.99	248.25
	Sick pay per day	£12.10			
	for 5 days =			60.50	60.50
Trade supervision	No. of tradesmen per foreman	8			
	Plus rate for foreman	£0.32			
	Time on supervision (%)	50		595.23	452.20
Working rule agreement	Tool money per hour	£0.05		93.52	
	Plus rate per hour (Labourer only)	£0.12			224.45
			Sub-total	10,569.58	8074.22
Overheads	Training levy (%)	0.25		26.42	20.19
	Employer's nat ins (%)	7		739.87	565.20
	Holidays with pay	19.6		921.20	921.20
	Public holidays			317.79	236.50
			Sub-total	12,574.86	9817.30
Severance pay and sundries	(%)	2		251.50	196.35
			Sub-total	12,826.36	10,013.65
Empl. liability and third party ins.	(%)	2		256.53	200.27
Annual cost of operative				13,082.89	10,213.92
Divide by total productive hours (1833) **Cost per hour =**				**£7.14**	**£5.57**

However, the establishment of realistic production standards is a major consideration outside the scope of this Code, but there can be no substitute for comprehensive company data and feedback from previous work of a similar nature to the project being priced.

Unit rates for measured items in the bills of quantity (excluding preliminaries) consist of any or all of the basic elements:

- Labour.

- Plant.

- Materials.

- Sub-contractors.

- Overheads (site and head office).

- Profit.

It is recommended that each element is analysed and estimated separately (as shown in the example in Figure 7.5) and that the total cost of each of the elements is considered by management. This approach is sometimes called 'analytical' estimating, which is compared with alternative techniques in Chapter 8.

Production standards and other considerations

The records of cost and outputs achieved on similar work from previous projects is a major source of information used in estimating. These data arise from records of resources used on site or from work study exercises to establish standards. It is important to remember that the cost or output depends upon many variables and attention should always be paid to the conditions which prevailed at the time when the particular recorded cost or output was noted and also to consider the levels of incentives which were used to achieve the particular standard. These conditions must be compared carefully with those expected to be encountered on the project under consideration. Differences between the estimated and actual cost or output on previous projects should be analysed and any obvious conclusions noted. Adjustments must then be made to update the estimating data.

When a particular type of work is being considered for the first time there will be no previous cost or output records for guidance. Trade specialists should be consulted whenever possible and technical information from outside sources may have to be used. Information can be provided by manufacturers either through printed literature or technical representatives. Caution will be needed when using data from external sources.

Proper allowances must be made for the 'learning curve' on the new type of work and for incentive payments, either by increasing the all-in labour rate or by using an appropriately modified production standard. It may be necessary to adjust for contractor's bonus which has been included in the all-in labour rate. This can be adjusted later in the project overheads, if necessary.

Labour element

Labour costs are estimated on the basis of the all-in hourly rates previously established. It is recommended that 'gang costs' should be used for some trades in preference to individual hourly rates. Typically an effective rate is built-up for a member of a concreting gang or a bricklayer rate will include a proportion of labourer's time. However, it is also reasonable to price all trades on a normal hourly rate basis providing an allowance for attendant ancillary labour is added to the established all-in hourly rates or added as an item of general labour in the project overheads stage.

It is usual to express outputs as 'decimal constants' such as 1.50 hrs to lay a square metre of brickwork This is because computer systems conventionally expect estimators to rate items by inputting a resource code and a quantity (see Figure 7.4).

Contractors and trade specialists assemble tables of data for use by their estimators as a guide to basic outputs. Figure 7.2 shows examples of constants for blocklaying.

Many factors affect the time allowed for an operation or item and careful consideration must be given to each of these factors enabling the time allowed to be as accurate as possible.

The drawings, specification and bills of quantity should be carefully examined to determine:

- The quantity of work to be done.

- The allowance needed for compaction, overbreak, batters, etc.

- Quality of finish and standard of workmanship required.

Figure 7.2 Typical data sheets with outputs for blockwork

| KEYSTONE | Estimating data | | **Lightweight blockwork** | | |
| KEYSTONE | Estimating data | | **Hollow blockwork** | | |

KEYSTONE	Estimating data		**Dense blockwork**		
	75	100	150	190	215
	Outputs hrs/m²				
Lay – lge quants	0.54	0.59	0.72	0.81	1.04
Lay – med quants	0.60	0.65	0.80	0.90	1.15
Lay – sml quants	0.72	0.78	0.96	1.08	1.38
Add for high walls	0.18	0.20	0.24	0.27	0.35
Add for filling openings	0.42	0.46	0.56	0.63	0.81
Add for dwarf walls	0.21	0.23	0.28	0.32	0.40
Add for casings	0.36	0.39	0.48	0.54	0.69
Mortar quants m³/m²	0.005	0.007	0.010	0.013	0.014

- Whether operations are repetitive.

- Whether excessive or detailed setting out will be required.

- Degree of accuracy and tolerances required.

- Whether the design of the work is intricate or straightforward.

- Whether any special skills will be needed.

- Whether any special construction sequence is necessary.

- Whether the operation is likely to be within the experience of existing staff and operatives, or whether special instruction or training will be needed, or whether there will be a need to engage specially trained personnel.

- Position on site in which the work occurs.

- Accessibility of work.

- Height or depth of work.

- Any double handling of materials.

- The weight of specific items.

- Restrictions in working, such as secure areas, safety.

- Shift times.

- Environment, such as hot/cold/exposed areas of work.

The tender programme and method statement will indicate:

- Time available for activities on the site.

- Time of the year when work is to be carried out and the likely seasonal conditions to be encountered.

- Whether work will be continuous or intermittent.

- Any restrictions which might affect normal working.

- Degree of interdependence of trades and operations.

- Facilities available for use by domestic and nominated sub-contractors as items of general attendance.

- Pattern of production and the likelihood of achieving maximum possible rates.

- Resources needed, such as the relative proportions of supervisory, skilled and unskilled operatives required, and recommended gang sizes.

- Extent of mechanisation envisaged and method of unloading, storing, handling and transporting materials.

The visit to the site and locality will have shown:

- Physical conditions and any restrictions likely to be encountered.

- Site layout, operating storage and unloading facilities.

- Likely skill, experience and availability of local labour.

The production standards to be used by the estimator in establishing labour costs will take into account the company's existing data and experience, and any circumstances associated with the project information, tender programme, method statement or visits to the site and consultants, which indicate that these standards should be modified in some way. The estimator must record any special factors which lead to alteration of production standards to any considerable extent, in the Estimator's Summary and Report for consideration at the final review stage.

Materials element

Various matters have now to be considered with respect to the quotation to be used in establishing the net unit rates. On page 64, Figure 5.4 was used to compare quotations for materials but other circumstances have also been taken into account. Consideration should be given and cost should be established for various additional matters associated with materials. These include:

- Any specific divergence or discrepancies from the contractor's enquiry in the quotations received from the supplier.

- Any minimum delivery requirements and adjustment of cost due to delivery in small quantities.

- Trade discounts which should be noted separately and reported at the final review meeting (*Note*: Discounts may or may not be deducted from the materials cost at this stage. Some contractors maintain that materials costs should be net of discounts, which are summarised in the estimate summaries. Others allow the discount to remain in the materials cost but recognise the element when considering the profit mark-up at the final review stage.)

- Waste allowances.

- Unloading, storage and distribution costs.

The degree of mechanisation in unloading must be considered to ensure that material deliveries are compatible with the intended method of handling; such as palleted materials for handling by forklift truck. Special equipment should be considered for unloading, although the costs of skips, slings and chains are more likely to be costed in project overheads than allocated directly against unit rates. The labour costs of unloading and distributing materials must be considered and an allowance made when establishing the total labour requirements of the project. Such labour can either be taken into account when selecting production standards for labour, or can be priced as a project overhead. Items to be accommodated include:

- Storage needs and protection.

- Size and weight of materials.

- The cost of any special packaging and crates, if these are charged, or the cost of returning them to the supplier.

- Any subsidiary fixing materials or temporary materials needed for storage.

The tender programme and method statement indicates:

- Time and rate of delivery required for materials.

- Amount to be stored on site, the location and method of subsequent distribution.

- Unloading point.

The visit to the site and locality shows:

- Physical conditions and any restrictions likely to be encountered.

- Site layout, operating, storage and unloading facilities.

The allowance made for waste must, wherever possible, be based on experience gained on previous projects. Data given in text books, periodicals and manufacturers' catalogues should be examined critically and used with caution. The waste allowance must be carefully applied according to the circumstances of the project and previous experience of the material.

The identification and costing of these various factors and considerations will convert the basic cost contained in the quotation into the cost which will be inserted into the net unit rates.

Plant element

The analysis of quotations received for plant will be set out in the 'Plant quotations register' in Figure 5.5 (see page 66) and any additional factors to be priced identified. Allowances must be made for additional matters associated with the plant. In considering the total costs of plant, decisions must be made concerning the following items:

- The manner in which time-related charges and fixed charges will be accommodated, i.e. delivery, erection and removal charges could be spread across the duration of hire and added to a weekly rate or alternatively shown separately as a fixed charge separate from time-related costs in the project overheads.

- Rate of production likely to be achieved by the plant, bearing in mind the specific requirements of the project, the season of the year, and in the case of excavation work, the ground and water conditions. In all cases, the tender programme requirements must be considered.

- The continuity which can be expected for any item of plant and the likelihood of achieving a high production rate should be evaluated. It is unlikely that outputs quoted by manufacturers can be attained.

- Average output should be established, making due allowance for intermittent working, site conditions, seasonal effects and maintenance.

In establishing the costs of plant, allowances must be made for additional matters associated with the plant. These include:

- Divergence or discrepancy from the contractor's enquiry in the quotation which is being considered.

- Delivery, erection and removal charges if applicable.

- Fuel costs, if applicable.

- Availability of power for electrically operated plant. Consider need for temporary sub-station or generators.

- The effect and cost of maintenance and consequent down time of plant.

- Special provisions needed for unloading and loading plant.

- Temporary access roads, hard-standings or temporary works required for the plant.

- Weight restrictions which may affect the plant or its use.

- Whether any special insurances are needed for the plant, such as responsibility for the plant during delivery and erection.

- Consents required for the use of plant on or over adjacent land.

- Contractor's attendant labour requirements, banksmen being particularly important.

- Safety measures that are required.

- Supporting equipment needed to operate plant, such as crane slings, chains, skips, cages, etc., associated with lifting equipment, hoses, breaker's points, etc., associated with a compressor – these items may be separately priced in the project overheads.

- Allowance for damage, repairs and replacement parts chargeable to the contractor.

- Minimum hire charges.

The estimator must decide which items will be accommodated in the all-in rate for plant, where plant is to be allocated against unit rates, and which items are to be allocated in the project overheads.

Allocation of costs

When mechanical plant is used only on specific and limited operations (such as excavation and soil disposal), there is little difficulty in allocating the costs of the plant to specific items measured in the bills of quantity, taking into account the various factors noted above.

However, when an item of mechanical plant serves a number of trades or operations (a crane or hoist or a concrete mixer, which is used for concrete work and also brickwork and drainage work), then the allocation of its cost to measured items can only be made on an arbitrary basis. When the cost of an item of plant is associated with time on site rather than to specific items of measured work (i.e. pumping operations), then such items cannot reasonably be allocated against measured work.

In such circumstances, the cost of such plant together with time in excess of productive output must be included in the project overheads, rather than spread in an arbitrary manner over measured rates.

There are many examples of resources which are difficult to allocate with one cost category. For example, falsework to support soffit formwork may be included in the material or plant element of a unit rate or equally can be assessed separately in project overheads as temporary works. For building work the cost of falsework is commonly in the net unit rate; but this is not the case for civil engineering where all supporting equipment and structures are linked to a resourced programme and temporary works calculations.

7.6 PRICING NOTES AND UNIT RATE CHECKLIST

Pricing notes

There are many techniques used by estimators in building-up unit rates, ranging from rough notes in the bills of quantity to sophisticated forms showing the constituents of rates broken down into rates and totals for labour, materials, plant and sub-contract elements. If there are changes made to the estimate at the final review meeting, the estimator's hand-written pricing notes could not be used (without further explanation) by construction staff except as a guide to the general logic and pricing structure.

A computer system, on the other hand, will enable an estimator to incorporate all the adjustments made to an estimate, and thus produce a valuable cost control document for site. The strengths of computer-aided estimating systems are the rapid re-calculation and reporting facilities. One drawback is the difficulty management may have in gaining an understanding of how rates were built up. Attempts to bring computers into the review meetings have tended to cause confusion and the danger that the estimate departs from the reports in front of management.

Unit rate checklist

A unit rate checklist can be used to remind the estimator of the main constituents of a unit rate. The example given in Figure 7.3 distinguishes between costs which are incorporated in the rate and those which could be priced in the project overheads.

For clarity, Figure 7.3 does not include a sub-contract element which could either be the whole rate or one of the constituents. Sub-contract rates are commonly added to rates for alterations and repairs such as 'remove floor coverings from main entrance and apply levelling compound'. In this example, the levelling compound work may be part of a flooring specialist's package.

Figure 7.3 Unit rate checklist

Category	Net unit rate	Project overheads
Labour	*Primary labour*	
	Prepare	
	Cut	
	Make	
	Fix	
	Adjust	
	Finish	
	Snagging	Snagging gang
	Assistant labour	
	Unload	Distribution gang
	Store	Distribution gang
	Distribute	Distribution gang
	Mix materials	
	Select materials	
	Clean up	Cleaners
Plant	Hand tools	Small tools allowance
	Production plant	
	General plant	Mechanical plant
	Power	Temporary services
	Fuel	
	Access equipment	Temporary works
Materials	Price quoted by supplier	
	Discounts	
	Delivery charges	
	Packing/pallet charges	
	Crane off-loading	Site cranage
	Overlap (for sheet materials)	
	Penetration (for aggregates)	
	Compaction	
	Conversion	
	Fixings and adhesives	
	Mortar and bedding materials	
	Sealants	
	Waste	
	Cutting from larger pieces	
	Residue from large packs	
	Breakages during fixing	
	Loss during fixing	
	Below specified standard	
	Theft and unclaimed losses	Contract conditions/insurances

7.7 DOMESTIC SUB-CONTRACTOR ELEMENT

The analysis of quotations received for domestic sub-contractors is set out in Figure 5.6, the 'Domestic sub-contractors' register' (see page 68). This analysis identifies any further matters which have to be costed by the contractor. Selection of the sub-contractor to be used in the net unit rates may not be possible before such additional costs have been determined.

Allowances must now be made for any additional matters associated with the domestic sub-contractor's works. These can include:

- Specific divergence or discrepancy from the contractor's enquiry included in the quotation.

- Allowance for unloading, storage, protection or materials and equipment and transfer of goods from stores to point of work, if this is to be the main contractor's responsibility. The labour costs associated with unloading and distribution of materials is considered at this time and allowance made when establishing the total labour requirements of the project. Such labour is either taken into account by an addition to the sub-contractor's quotation or can be priced in the project overheads.

- General attendance items to be provided by the main contractor (see Chapter 10).

In making such allowances, the contractor must take into account the requirements of the tender programme and method statement and facilities which have already been allocated for the contractor's own works. Additions made to cover attendance on domestic sub-contractors may be done in several ways, by:

- Increasing the relevant unit rates of the sub-contracted work.

- Adding a fixed percentage to the whole of the sub-contractor's quotation.

- Making an addition subsequently in the project overheads.

Discounts offered by sub-contractors must be noted separately and reported at the final review meeting. (*Note*: Discounts may or may not be deducted from the sub-contractor's quotation at this stage. Some contractors maintain that sub-contractor's costs should be net of discounts, which are summarised in the summary reports. Others allow the discounts to remain in the cost of the work to be sub-contracted but recognise the element when considering the profit mark-up at final review stage.)

Great care must be taken in assessing sub-contract quotations to ensure that all items have been adequately covered. If labour-only sub-contractors are being considered, the cost allowance *must take into account all factors associated with the provision of materials by the contractor*, and adequate safeguards must be made to control the use and wastage of such materials.

7.8   COMPUTER METHODS FOR UNIT RATE PRICING

Computers are an established part of an estimating office for all sizes of organisation and bring benefits in the ways in which estimates can be controlled and reports generated.

The technique of resourcing a bill item (pricing) follows certain rules whether done manually or using a computer. Computer systems generally restrict the input of resources to a standard format with the following features:

- A resource can be identified by finding it in a look-up window, or by entering a code which brings the resource to the screen.

- The resource name can be created or changed.

- The unit for the resource can be modified.

- A quantity can be entered for the resource (such as 1.25 if a sheet material requires an additional 25 per cent for laps).

- A factor or wastage allowance can be applied to the resource quantity.

- The cost of a resource can be added or modified.

An example of a rate built up using a computer is given in Figure 7.4. This shows that outputs are expressed as decimal constants and a limited amount of calculation is possible in analysing a resource. A special kind of resource can have a number of sub-resources within it. They are called 'activity' or 'gang' resources and enhance the flexibility of a system. The use of activities reduces the number of resources which the estimator must enter for certain items of work. In Figure 7.4 the excavator includes plant operator and banksman, both being part of the excavator resource.

Figure 7.4 Computer pricing system

Bidwell Estimating Systems   Tender: Helix Laboratories Limited

Cost code	Item ref.	Description	Quant.	Unit	Rate	Total
D20	3/1a	Excavating	245.00	m³	5.60	1372.00
		Trenches exceeding 300 wide				
		1 m maximum depth				
		commencing 600 below existing ground level				

Resource code	Resource description	Unit	Quant.	Resource rate	Factor/ waste	Unit rate
D20EXC	EXCAVATOR AND BANKSMAN	HR	0.20	22.00	1.00	4.40
D20LAB	GROUNDWORKS LABOURER	HR	0.20	6.00	1.00	1.20
						5.60

In order to gain the benefits of computer systems, estimators must accept the discipline of a structured approach to data entry and give up the flexibility that has been possible with manual methods. For the example given above a typical manual calculation is shown in Figure 7.5.

7.9 PRICING CHECKS

When management considers risks at the final review meeting, an assessment is made of commercial and technical matters and it is assumed that the estimate has been correctly calculated with very few errors. In order to eliminate significant calculation errors an estimating department needs procedures for ensuring that standards are maintained. The objectives of a quality system are dealt with in Chapter 11.

A checklist is given in Chapter 4 for monitoring the input of bill items using a computer system. Once the items and quantities have been entered correctly, the estimator is normally responsible for accurately building up unit rates, and his work is checked in four ways:

1. Forms can be designed to add across and downwards so that a page total comes from two directions. Examples of self-checking forms can be seen in the project overheads summaries.

Figure 7.5 Manual pricing system

Item details				Analysis				Totals
Ref.	Description	Quant.	Unit	Lab.	Plant	Mat.	Sub.	
3/1a	Excavator 5m^3/hr @ £16.00/hr				3.20			3.20
	Banksman 5m^3/hr @ £6.00/hr			1.20				1.20
	Labourer trimming 5m^3/hr			1.20				1.20
	Unit rate	1	m^3	2.40	3.20			5.60
		245	m^3	588.00	784.00			1372.00

KEYSTONE CONSTRUCTION LTD — PRICING NOTES
Trade: Earthworks
Tender number
Page no.

2. By questioning during the review meeting, an estimator will explain how the principal rates were calculated.

3. The preparation of summaries of resources enables the estimator to list the most significant items and management can assess the discounts, wastage factors and quantities allowed in the estimate. It is important that this reconciliation is made without relying entirely on the reports generated by a computer system. Examples of such forms are given in Figures 11.2 to 11.5 (see pages 156 to 159).

4. Quotation totals from suppliers and sub-contractors can be checked against totals incorporated in the estimate.

Where computers are not used, unit rates are written into a copy of the bill of quantity and clerical assistants extend the rates and add page totals in order to arrive at a total for the project. In this case, verification of the calculations is carried out by a second 'checker' to ensure the total is correct.

For tenders based on drawings and specifications, an independent check on the principal quantities must be carried out in order to reduce the risk of a mistake, either leading to the winning of an undervalued bid or producing a high tender which may influence a client in giving opportunities to bid for future projects.

8. ALTERNATIVE ESTIMATING TECHNIQUES

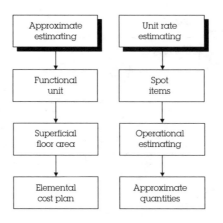

8.1 INTRODUCTION

This Code deals with the production of an estimate and tender where a detailed design and a bill of quantity is available, and the constituents of rates (labour, plant, materials and sub-contracts) are priced separately and summarised separately for the review meeting and handover. With the increasing use of computers for estimating, this is by far the most common way to produce a competitive estimate.

At an early stage in a project's development, approximate estimating techniques are employed in order to set budgets and assess the feasibility of a scheme and, where necessary, gain funding. During design development a cost plan is an important way of producing a design within the client's budget.

8.2 APPROXIMATE ESTIMATING

Approximate estimating techniques depend on historical cost data being available from previous similar schemes. In practice this information is analysed and applied in three ways:

- Functional unit or unit of accommodation method.

- Superficial (floor area) method.

- Elemental cost plan.

Functional unit or unit of accommodation method

This method uses the unit cost from earlier schemes. The unit cost is calculated by dividing the total cost of building by the number of functional units. For example, the functional unit for a motorway hotel might be a single room, the cost of which includes a small proportion of the front office and corridor space. A hospital bed may be the basic unit used in a general hospital and a car space in a multi-storey car park. Clearly this approach can be used to set broad yardsticks in setting budgets for construction but it must be recognised that there is a significant difference between this budget and the final cost of construction. This is because the nature of the site varies, the client's brief has to be developed, incoming services are not always available close to the building, the shape and specification for the building affects the price as will commercial aspects of the contract when the project is tendered.

Superficial (floor area) method

This method uses historical data from earlier comparable schemes in terms of the cost per square metre of floor space. This is a popular method readily understood by developers and the building team, as relatively few rules are needed to apply this technique. The floor area of a building is defined as that measured at each floor level using the internal dimensions, without making deductions for internal walls or stairs. As with any approximate estimating technique, a number of adjustments are needed to take account of location, specification, degree of complexity, size, shape, ground conditions and number of storeys.

In order to assess these factors, reliable historical costs are needed from a variety of buildings within each building category. A separate assessment must be made of external works, main services and drainage which can all vary substantially depending on the nature and location of the site.

Elemental cost plan

An *elemental cost plan* can be produced from a preliminary building design. Again the method depends on reliable data being available from comparable projects where the actual costs for each building element is known. Figure 8.1 shows how prices from a bill of quantity for a similar office building have been set against a standard list of elements. In this example, the floor area of the previous building was

Figure 8.1 Elemental cost plan

KEYSTONE CONSTRUCTION LTD	**ELEMENTAL COST PLAN**		Project Helix Laboratories Type Two-storey offices	

	KB Electronics Gross floor area = 2900 m^2 Elemental costs			Helix Laboratories Gross floor area = 3850 m^2		
	£/m^2	Totals	Notes	£/m^2	Totals	Notes
Substructure	68.10	197,500	RC pads	95	365,750	piled founds
Steel frame	68.81	193,750		67	257,950	
Upper floors	36.21	105,000		36	138,600	
Roof	50.43	146,250		50	192,500	
Stairs	14.22	41,250		14	53,900	
External walls	62.50	181,250		63	242,550	
Windows and external doors	46.55	135,000		47	180,950	
Internal walls and partitions	31.03	90,000		31	119,350	
Internal doors	6.59	19,125		7	26,950	
Wall finishes	13.36	38,750		13	50,050	
Floor finishes	61.21	177,500		61	234,850	
Ceiling finishes	17.67	51,250		18	69,300	
Steelwork (fireproofing)	3.88	11,250		4	15,400	
Fittings and furniture	9.48	27,500		9	34,650	
Sanitary appliances	5.60	16,250		6	23,100	
Disposal installations	4.40	12,750		4	15,400	
Water installations	6.03	17,500		6	23,100	
Heat source and space heating	52.16	151,250		52	200,200	
Ventilation and cooling	20.69	60,000		21	80,850	
Electrical installation	103.45	300,000		103	396,550	
Lift installation	0.00	0		13	50,050	two lifts
Security alarms	27.28	79,125		27	103,950	
Fire alarms	10.73	31,125		11	42,350	
Builder's work in connection	7.03	20,375		7	26,950	
Minor works	3.10	9,000		3	11,550	
Total elemental cost	728.53	**2,112,750**		768	**2,956,800**	
Preliminaries for building	58.31	169,100		58	223,300	
Net building cost	728.53	**2,112,750**		768	**2,956,800**	
Site works	105.39	305,625		105	404,250	
Drainage	24.18	70,125		24	92,400	
External services	7.16	20,750		7	26,950	
Preliminaries for ext works	8.21	23,800		8	30,800	
Net scheme cost	873.47	**2,533,050**		912	**3,511,200**	
Design fees	44.83	130,000		45	173,250	
Statutory fees	11.03	32,000		11	42,350	
Overheads and profit	55.17	160,000		55	211,750	
Contingencies	34.48	100,000		35	134,750	
Budget total	873.47	**2,533,050**		912	**3,511,200**	

$2900\,\mathrm{m}^2$. It is shown that the breakdown of costs for a $3850\,\mathrm{m}^2$ building can be calculated simply by applying the earlier proportional cost in each element to the second building. A number of adjustments are easily made; such as, the introduction of a lift and a piled foundation.

8.3   UNIT RATE ESTIMATING

A large part of an estimator's time is devoted to calculating unit rates for items in a bill of quantity. In addition to analytical unit rate pricing, which is dealt with in Chapter 7, the estimator will use various techniques.

Spot items

These are operations that are difficult to break down into discrete items of work in a bill of quantity. For example, the demolition of small buildings or forming openings through walls are both priced by looking at the extent of the work during a site visit.

For estimating purposes, spot items may be treated in several ways:

- Approximate quantities can be taken off and unit rates used to calculate a lump sum estimate for the item.

- The description within the bill can be analysed into its constituent operations and trades and an estimate of the cost made for each.

- When the description within the bill is analysed into constituent operations and found to have a predominant trade, then a gang or operational assessment can be made on a time, plant and material basis so that the overall cost can be calculated.

In all three methods the cost of labour, plant and materials should be separated in accordance with the general principle described for unit rates in Chapter 7 (see page 80).

The work must be inspected thoroughly at the site visit and, where necessary, the construction method must be established and documented. Adequate allowances must be made for storage, temporary work, including supports, access, double handling, small deliveries, making good and reinstatement. Transport can be included in the item but will usually be included in project overheads.

If bill descriptions are not clear, or if further information or measurements are required, it may be necessary to re-visit the site.

Operational estimating

This system is adopted when the estimator needs to consider the overall duration of an operation and its interrelationships with other trades. This is the case with civil engineering construction or the earthworks and concrete elements of a building project. In these cases, it is unrealistic to look at a single unit of work and wrong to assume the total cost of an operation is the product of the unit rate and quantity.

For example, a contractor may make an assessment for laying precast concrete manhole rings on the basis of the number a drainage gang can fix in one day. For a 2.10 m diameter manhole and 600 mm high units, the estimator might assume that 15 units can be handled, lifted into position and securely bedded in one day. If a project has 25 precast concrete units, an allowance of two days may be needed because it might be difficult to deploy the plant elsewhere for a small part of the second day. It is unlikely that a simple unit rating exercise would have included an allowance in this way for standing time.

Operational estimating depends on a careful study of how a section of work will be carried out in practice. It is difficult, for example, to price the fixing of roof trusses without looking at working methods.

Case study 1

A new dental surgery has a rectangular plan shape, 35 m long, with 55 nr timber roof trusses above first floor level spanning 8.50 m between wallplate supports. The estimator has drawn up a list of resources for fixing trusses as follows:

Mobile crane (1 nr)	1 day	@	350.00 =	350.00
Banksman (1 nr)	8 hrs	@	7.00 =	56.00
Carpenters (2 nr)	16 hrs	@	9.00 =	144.00
	for 55 trusses			£550.00
	for 1 truss			£10.00

In this case a method statement was not produced. But for more complex construction operations, for example, more planning is needed together with method statements.

Case study 2

A large distribution warehouse has a floor slab with 850 m³ of concrete to be cast in continuous pours with all joints and reinforcement to be introduced during the casting operation. For the method chosen 300 m³ of concrete will be placed each day.

After a detailed review of the design and discussions with the consultants it was agreed to replace steel reinforcement with fibre-reinforced concrete. The estimator and planning engineer drew up a list of resources and durations, as follows:

Trades foreman	(1 nr)	3 days × 9 hrs @ £12.00	324.00
Gangers	(2 nr)	3 days × 9 hrs @ £10.00	540.00
Operatives	(10 nr)	3 days × 9 hrs @ £8.00	2160.00
Carpenters	(2 nr)	3 days × 9 hrs @ £9.00	486.00
Laser screeder	(2 nr)	3 days × 9 hrs @ £10.00	270.00
Drivers	(1 nr)	3 days × 9 hrs @ £9.00	243.00
Dumper	(2 nr)	3 days × 9 hrs @ £3.00	162.00
Drivers	(2 nr)	3 days × 9 hrs @ £9.00	486.00
Power floats	(2 nr)	3 days × 9 hrs @ £3.00	162.00

Total cost for 850 m^3 £4833.00

In order to insert a unit rate in the bill of quantity, the total cost for placing concrete is divided by the quantity.

Unit rate for labour and plant = £4833/850 m^3 = £5.69/m^3

It is not difficult to use computer-aided estimating software for operational estimating. A list of resources can be produced for an item of work and applied to the total quantity, the computer automatically divides the total cost by the quantity in order to arrive at a unit rate. The last example would be input in the following way:

Concrete slab = 850 m^3
Pricing quantity = 850 m^3
Total cost = £4833.00
Unit rate = £5.69

Code	Description	Unit	Quantity	Rate
LFT	Trades foreman	(hr)	27.00	12.00
LGA	Gangers	(hr)	54.00	10.00
etc.				

Approximate quantities

Approximate quantities are needed where the other approximate estimating techniques do not produce sufficient information for a reliable budget. The term 'approximate quantities' is used for a number of arrangements, the most common being:

- A shorter bill of quantity with composite items. An item for external walls, for example, would include both skins of masonry, forming the cavity, wall ties, plastering and pointing. In this case an approximate bill of quantity is produced and priced with rates taken from a number of sources including previous bills of quantity, price books or guide prices from specialist trade contractors and suppliers. The accuracy of this method will depend on the extent to which the design has been developed. If approximate bills of quantity are to be used for cost planning during the design stage, they should follow an elemental bill format giving estimated costs for each building element.

- A contractor's bill of quantity produced from drawings and specifications will include fewer ancillary items which are required by the standard method of measurement. The rules for measurement, such as deducting openings in walls, are often ignored, as it is assumed that the over-measure will account for the extra labour and increased waste on materials. Computer-aided estimating systems provide a fast method for creating a bill of quantity. The estimator selects items from a library of descriptions which were previously priced. Resource costs can be changed when material quotations are received.

- In order to start a contract early, a bill of quantity (often from another project) can be used to establish a tender price. The JCT Standard Form of Building Contract has a variant for this arrangement.

9. PC AND PROVISIONAL SUMS

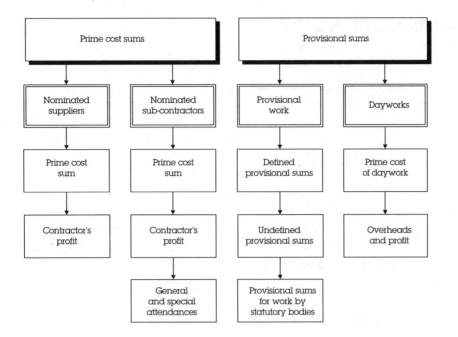

9.1 INTRODUCTION

The contents of tender submission documents vary considerably depending on the form of contract and documents used to define the works. Bills of quantity for building work are divided into five sections:

- Preliminaries.

- Preambles (specification).

- Measured work.

- Prime cost and provisional sums.

- Summary.

In order to avoid confusion in calculating and analysing an estimate, it is recommended that prime cost and provisional sums should form a separate section at the end of the measured work part of a pricing document. Since this is not always the case estimators must carefully check all the bill pages, including preliminaries, to ensure that the written-in sums are incorporated in the final tender amount.

9.2 NOMINATED SUPPLIERS

Standard methods of measurement state that the cost of materials from nominated suppliers are to be identified in the tender documents as prime cost sums. A separate item is also given for the contractor to add his profit. PC sums may also be written-in to an item description (such as a rate for the supply of facing bricks) for the estimator to incorporate the cost in his rate build-up.

The estimator will produce a list of nominated suppliers at an early stage using the 'Schedule of PC sums and attendances' (see Figure 9.1). Where details associated with a nominated supplier are unclear, the estimator must note any concerns in his report for further consideration at the final review meeting.

If a PC sum has been included for high-value materials or large quantities the estimator must check the following:

- The terms of the purchase contract – JCT '80 provides for the nominated supplier to allow the contractor a discount (for payment in full within 30 days of the end of the month during which delivery is made) of 5 per cent. This discount is normally deducted from the estimate in order to include net costs in the summaries for the final review meeting. Since this discount is not available through the terms of all main contracts, the estimator must check the relevant contract conditions and not assume a discount is available. The estimator must also assess the risk which would arise if the final value of PC sums for nominated suppliers is not achieved and too much discount may be deducted from the value of the contract at tender stage.

- Delivery times and their affect on the programme.

- Fixing items associated with materials provided by a nominated supplier need to be adequately described and measured in the items to be priced. Any discrepancies concerning fixings, such as bolts, screws, brackets, adhesives and sealants, or ambiguity over the responsibility for supply of these items, must be clarified.

- Additional costs for unpacking, storing, handling, hoisting and the return of re-usable crates or pallets to the supplier. Suppliers may deliver their materials in re-usable crates or other packaging and the contractor may be required to return such items to the supplier. Due allowance must be made for the collection, storage, handling and subsequent dispatch of such items back to the supplier.

Where bills of quantity are used, the fixing of materials supplied by nominated suppliers is measured in the appropriate part of the bill.

9.3   NOMINATED SUB-CONTRACTORS

A nomination arises in construction contracts where the selection of a sub-contractor is to be made by the client or his representative, for which a prime cost sum has been inserted in the tender documents.

There is a right of objection to a particular nominated sub-contractor because it would be contrary to contract law if a party is forced into a contract unwillingly. Where prime cost sums are included in bills of quantity, the estimator is seldom given the name of the proposed sub-contractor and so it is impossible to discuss methods and programming issues.

The use of nominated sub-contractors is becoming increasingly rare for two reasons:

- The growing complexity of nomination procedures.

- The added risks carried by a client when a proportion of the responsibility for full performance is taken from the contractor.

Where nominations are used they can account for a significant proportion of the overall cost of a contract. Contractors are frequently given inadequate supporting information to deal with attendances in the bills of quantity. The adequacy of the information provided must be carefully investigated, and further particulars requested by the estimator if details are not complete.

Adequacy of information

The standard method of measurement gives the items to be included in the bills of quantity for each nominated sub-contractor, as follows:

- The nature and construction of the work.

- A statement of how and where the work is to be fixed.

- Quantities which indicate the scope of the work.

- Any employer's limitations affecting the method or timing of the works.

- A prime cost sum.

- General attendance item.

- An item for main contractor's profit, to be shown as a percentage.

- Details of special attendance required by the sub-contractor.

At tender stage, the estimator must check that the measured items for works which are covered by a PC sum are adequate and that supporting details are available in accordance with the appropriate standard method of measurement. If not it will be necessary to ask for further particulars before the interrelationships of trades, the tender programme and method statement can be finalised. All too often this is a neglected area and the estimator must ensure that the tender programme reflects sufficient time for the execution of work which is the subject of a PC sum and that all trades are effectively co-ordinated.

Clarification must be obtained from consultants if there are any doubts as to the adequacy or meaning of the descriptions used.

Attendance

Attendance is defined as, *'the labour, plant, materials or other facilities provided by the main contractor for the benefit of the sub-contractor and for which the sub-contractor normally bears no cost'*. The main contractor is responsible under the main contract provisions for the site establishment and providing attendance. This provides clear responsibilities for the support services and equipment needed on site and eliminates duplication of resources for various specialist sub-contractors. For very large contracts, where a construction manager or management contractor has overall control, trade contractors are asked to provide certain parts of the temporary works and facilities themselves. The costs associated with attendance are built into the main contractor's tender and consequently become a charge against the client. However, the associated risks of attendance are borne by the main contractor.

The estimator must decide how to price 'general attendance' and 'special attendance' relating to nominated sub-contractors. The attendances may be priced in the project overheads schedules or on the 'Schedule of PC sums and attendances' form (see Figure 9.1).

KEYSTONE CONSTRUCTION LTD	SCHEDULE OF PC SUMS AND ATTENDANCES	Project Helix Laboratories, Westfield

PC sums

Bill page	Item	Nominated suppliers	Gross	Discount	Net
5/3	a	Ironmongery	5000	250	4750
	b	Doors	11,000	550	10,450
	c	Kitchen appliances	3500	175	3325
		Totals to summary	19,500	975	18,525

Bill page	Item	Nominated sub-contractors	Gross	Discount	Net
5/4	a	Structural steelwork	45,000	1125	43,875
	b	Lift installation	24,000	600	23,400
	c	Aluminium windows	74,000	1850	72,150
	g	Access road for steelwork			
		Totals to summary	143,000	3575	139,425

Attendances

Lab	Plt	Mat	Sub
600	600	1600	
600	600	1600	

Figure 9.1 Schedule of PC sums and attendances – Form F

General attendance

The item for general attendance is an indication of the facilities which are normally available to sub-contractors where they are provided by the contractor to meet his own requirements.

In assessing any sums to be allowed for general attendance, the estimator must investigate the facilities which will already be provided for the main contractor's use and determine any costs which may arise by the nominated sub-contractor's use of any such facilities. The facilities given in SMM7 include:

- Use of temporary roads, pavings and paths.

- Use of standing scaffolding.

- Use of standing power-operated hoisting plant.

- Use of mess rooms, sanitary accommodation and welfare facilities.

- Provision of temporary lighting and water supplies.

- Providing space for sub-contractor's own office accommodation and for storage of his plant and materials.

- Clearing away rubbish.

Use of temporary roads, pavings and slabs
Allowance must be made for any costs associated with the maintenance of temporary roads, pavings and paths which are required during the time period allowed by the contractor for his own use. This item will not cover any specific access requirements of a nominated sub-contractor. Such items, for example, as hard standing for a crane should be separately described under 'special attendance'.

Use of standing scaffolding
The contractor must allow for any costs which might arise through the nominated sub-contractor's use of scaffolding which is already erected for the main contractor's use. Any modifications or additional scaffolding required or any extension of programmed time for such scaffolding over and above the time period required by the main contractor must be described and measured as 'special attendance' in the bills of quantity.

Use of standing power-operated hoisting plant
While nominated sub-contractors may use existing hoisting plant if there is spare capacity, any hoisting facilities specifically required must be measured under 'special attendance'.

Use of mess rooms, sanitary accommodation and welfare facilities
Assessment must be made of the accommodation needed for the operations of nominated sub-contractors over and above the requirements of the contractor. Allowance must also be made for any servicing and cleaning of such facilities which are shared with the contractor.

Provision of temporary lighting and water supplies
The estimator must establish requirements for general lighting needed to comply with safety requirements and for the execution of the works during normal working hours. Adequate allowance must also be made for water points needed for the construction of the works. This may mean the simultaneous provision of such services in other areas of the building over and above the requirements of the main contractor.

Special lighting requirements and power needs must be measured and priced under 'special attendance'. Specific water requirements for testing or associated with commissioning of plant should be measured under 'special attendance'.

Providing space for sub-contractor's own office accommodation and for storage of his plant and materials
The estimator should note that only space is required and that cover in the form of a shed is not a requirement. The assessment of total space requirements must be borne in mind when finalising the method statement and site layout.

Clearing away rubbish
The disposal of waste, packaging and other rubbish from an agreed collection point involving labour, containers and haulage must be assessed. Abnormal items of rubbish, such as disposing of surplus excavated material from a ground improvement technique, must be measured separately under 'special attendance'.

Special attendance
Other specific attendances which do not fall under the category of 'general attendance' must be specifically measured in the bill of quantity as 'special attendance'. Items to be measured include:

- Special scaffolding or scaffolding additional to the contractor's standing scaffolding.

- The provision of temporary access roads and hardstandings in connection with structural steelwork, precast concrete components, piling, heavy items of plant and the like.

- Unloading, distributing, hoisting and placing in position, giving in the case of significant items the weight, location and size.

- The provision of covered storage and accommodation, including lighting and power thereto.

- Power supplies giving the maximum load.

- Any other attendance not included in 'general attendance' or listed above.

Special scaffolding
In order to price this item, the estimator must be given precise details concerning the scaffolding requirements. Such information should define clearly the height in stages of the scaffolding, indicate the extent of boarded platforms and any alteration and adaptation that will be required. If such information is not available and descriptions are inadequate, the estimator should seek further instructions from the consultant.

The estimator must also make due allowance under this heading for any adaption or alteration to standing scaffolding or for any extension to the time period, providing such items are described and measured in the bills of quantity.

The provision of temporary access roads and hardstandings
Where any specific requirements are described, these must be taken into account with the contractor's own needs and any additional temporary provision allowed for.

Unloading, distributing, hoisting and placing in position
This item may also include some intermediate storage requirement and due allowance should be made for this. It is essential that particulars are stated of the size or weight of materials to be handled to enable the estimator to reasonably assess costs and identify the appropriate mechanical aids. In the case of heavy units, e.g. precast cladding, it will also be necessary to be informed of the delivery rate and also if any specific stacking facilities are required for storage on site.

Sufficient information must also be provided to identify any distribution requirements, as opposed to hoisting and stacking. In assessing the cost of such specific facilities, the estimator must bear in mind the extent of use of existing mechanical hoists and must ensure that sufficient hoisting capacity is available to meet all needs.

In some cases the estimator needs to seek clarification if components such as precast concrete cladding panels or mechanical plant are to be placed in position. This is to prevent any overlapping of responsibilities with the sub-contractor who would be expected to supply and fix all his own materials.

Provision of covered storage and accommodation including lighting and power

Under 'general attendance' the contractor is required to provide space for nominated sub-contractors to erect their own facilities. Under this item the main contractor will be required to provide, erect and maintain accommodation and provide lighting and power as stipulated. The size of huts required should be stipulated and the period required stated. Any special requirements, i.e. racking or other services, should also be defined.

Power supplies giving the maximum load

Any special power requirements, including power for testing of systems, must be clearly measured for pricing purposes. Any reference to power supplies should state whether single- or three-phase supplies are required and the maximum demand level should be taken into account. The estimator should ensure that any descriptions for fuel or power for such testing purposes are clearly specified, giving the quantity necessary to fulfil the tests and also the precise specification of the power needs.

Maintenance of specific temperature or humidity levels

Any specific requirements for controlling temperature or humidity must be clearly measured, stating temperature/humidity required and the time period that the contractor must provide these services. The requirement must also state if the permanent services in the building can be used for this purpose.

Any other attendances

Under this heading, the contractor is required to provide specific attendance or materials for various trades. This could include the provision of bedding material for roof tiles, or floor tiles. Other items such as specific cleaning operations, and the removal of masking tape used by sub-contractors should also be defined. Such items must be clearly measured and, in the event of any inadequacies or ambiguities, the estimator should refer to the consultants for further instructions.

Builder's work

Items of builder's work in connection with works by nominated sub-contractors will be measured in accordance with the requirements of the Standard Method of Measurement. The estimator will then be required to establish unit rates for the measured items in accordance with the principles previously described. The following points should be borne in mind:

- The extent to which builder's work items are shown on the tender drawings. It may be necessary to ask for further particulars before pricing can be completed.

- The requirement for any specialist work or plant to carry out the builder's work required.

- The rate and timing at which the builder's work should be carried out.

- The continuity to be expected while carrying out the builder's work.

9.4 PROVISIONAL SUMS

Provisional sums are included in bills of quantity for items of work which cannot be fully described or measured in accordance with the rules of the method of measurement at the time of tender.

For work measured under the rules of SMM7, there are three types of provisional sum:

Provisional sums for defined works

This provisional sum is used where works are known to be required in the project but have not been fully designed or specified at tender stage, and so cannot be measured in detail. The contractor must make due allowance for the programming, planning and pricing preliminaries; and to enable him to do so the following information must be provided with the provisional sum:

- The nature of the work.

- How and where it is to be fixed.

- Quantities showing the scope and extent of the work.

- Limitations on method, sequence and timing.

As an example, a provisional sum for defined work would be stated as:

Allow the provisional sum of £8000 for the reception counter to be located in zone G3. The counter will be 'L'-shaped on plan, approximately 6.00 m girth and incorporate four workstations, and is to be installed after completion of floor finishes.

Provisional sums for undefined works

Where the information required in support of a defined provisional sum is not available the provisional sum is 'undefined' and the contractor is not required to take account of the costs of programming, planning or preliminaries.

Undefined provisional sums are typically used to make provision for possible expenditure on elements of work which cannot be wholly foreseen at tender stage or cannot be quantified. A client's contingency sum is deemed to be an undefined provisional sum.

As an example, a provisional sum for undefined work would be stated as:

Allow the provisional sum of £750 for remedial work to roof boarding exposed after the removal of existing roof coverings.

Provisional sums for works by statutory authorities

SMM7 makes provision for a provisional sum to be included in a bill of quantity which is neither 'defined' nor 'undefined', for work to be carried out by the local authority or statutory undertakings including privatised services authorities carrying out statutory works.

Incorporating provisional sums in an estimate

Although not strictly required by SMM7, it is common for items to be measured to enable the contractor to price attendance and profit on works by statutory authorities as though they were nominated sub-contractors. It should be noted that these provisional sums are net and do not include a contractor's discount.

With the exception of provisional sums for work by statutory authorities, provisional sums are deemed to include an allowance for main contractor's head office overheads and profit. All provisional sums will become the subject of an Architect's instruction during construction and the work will be valued according to the appropriate contract rules for measurement and valuation which include provision for overheads and profit. Accordingly, the value of provisional sums (except those for statutory authorities) should be added into the final summary after the application of overheads and profit. Alternatively, if it is company practice to add head office overheads and profit to the total value of measured and unmeasured work, provisional sums should be discounted before entry in the summary to avoid duplicating the overheads and profit for this type of work.

Figure 9.2 illustrates a typical schedule of provisional sums and daywork. The total for each category is carried forward to the final summary for review by management.

9.5 DAYWORK

Definition

Contractors must understand the circumstances in which varied or additional work will be valued on a daywork basis. It normally occurs where variations cannot be valued by measurement using bill rates or comparable rates; nor by negotiation before an instruction is issued.

The daywork charges are usually calculated using the definitions for prime costs and overheads published by the RICS/BEC for building work and FCEC for civil engineering.

The prime cost of daywork can be defined in other ways, and care must be exercised in reading the definition in the tender documents.

The composition of the total daywork charge will include the following costs:

- Labour.

- Materials and goods.

- Plant.

KEYSTONE CONSTRUCTION LTD	SCHEDULE OF PROVISIONAL SUMS AND DAYWORK	Project — Helix Laboratories, Westfield

Provisional sums

Bill page	Item	Defined provisional sums	£
5/6	a	Security fencing to rear yard	5000
	b	Reception desk	11,000
	c	Kitchen appliances	3500
		Totals defined provisional sums	19,500

Bill page	Item	Undefined provisional sums	£
5/8	a	Entrance gates	9000
	b	Additional insulation to roof space	2500
	c	Machine bases in plant room	4000
	d	Additional builder's work	4000
5/9	a–g	Various	21,000
5/11	d	Contingencies	36,000
		Totals undefined provisional sums	76,500
		Total provisional sums to summary	96,000

Dayworks

Bill page	Item	Type	Basic sum	%	Add	Total
5/12	a	Building labour	8000	100	8000	16,000
	c	Specialist labour	8000	120	9600	17,600
		Labour totals	16,000		17,600	33,600
5/13	a	Building materials	8000	15	1200	9200
	c	Specialist materials	8000	15	1200	9200
		Materials totals	16,000		2400	18,400
5/14	a	Building plant	4000	10	400	4400
	b	Specialist plant	4000	10	400	4400
		Plant totals	8000		800	8800
		Daywork totals	40,000		20,800	60,800

Figure 9.2 Schedule of provisional sums and daywork – Form J

- Supplementary charges (civil engineering contracts).

- Incidental costs, overheads and profit (this addition will vary between labour, plant and materials and, in order to introduce competition at tender stage, is added to provisional sums for the prime cost of labour in the bills of quantity by the contractor).

An alternative method (for labour to be valued on a daywork basis) is for the contractor to provide all-in gross hourly rates which are applied to provisional hours. This makes the calculation of daywork rates simpler during the course of the project but moves the burden for anticipating increased costs to the contractor.

Contractors may decide that some of the project and head office overheads are covered in the contract price and may be excluded from daywork rates. This is mainly true if the daywork to be carried out during the term of the contract will not result in extension to the contract, but other additional costs to project and head office overheads may still have to be considered.

It is inappropriate to use this payment method for anything except work which is incidental to contract work. In the event that significant changes are made to the original scope of works the valuation rules normally allow additional overhead costs to be recovered, usually when the full effects of changes are known.

Decisions concerning allowances for profit and overheads must be made by each contractor taking into account his own circumstances, method of working and his assessment of the effects of daywork on a particular project. The contractor must assess each contract on its own merits in producing daywork rates and calculating the percentage addition needed. This will include an assessment of the likelihood of the prime cost being a reasonable pre-estimate of the work which will be valued on a daywork basis.

The contractor's daywork percentages must take into account the rates required by the sub-contractors used in the tender. Enquiries to sub-contractors must include a request for daywork percentages based on the definition incorporated in the main contract. For mechanical and electrical installations in building contracts, the contractor is given the facility to state different percentages for specialists in the bills of quantity.

As dayworks are calculated inclusive of an allowance for overheads and profit, they should, like provisional sums, be added into the final summary after the application of overheads and profit.

Labour

The calculation for the prime cost of labour differs somewhat from that of the all-in rate; various incidental costs, overheads and profit are deemed to be included in the percentage added to the prime cost by the contractor. The percentage added must include all other items that the contractor considers are necessary to recover the true cost incurred but not allowed under the definition of prime cost of daywork.

The prime cost of labour calculation consists of:

- Guaranteed minimum wages and emoluments.

- Additional emoluments in respect of the Working Rule Agreement.

- Overhead costs in employing operatives.

- Incidental costs, overheads and profit – the percentage addition.

Calculation of percentage to be added to prime cost of labour

To calculate the percentage to add to the prime cost for daywork for labour the estimator must find the difference between the total cost of labour and the prime cost calculated using the definition.

Figure 9.3 uses the BEC/RICS *Definition of Prime Cost of Daywork* to compare the prime cost calculation for labour with the all-in rate produced in Chapter 7. The calculation shows that in this example £2.86 (51 per cent) has to be added to the prime cost of labour for incidental costs, overheads and profit. This is the percentage to be inserted in the bills of quantity.

In practice, contractors often add considerably higher percentages than this for the following reasons:

- To provide for additional supervision and record keeping.

- To account for the disruption to the programme which might be difficult to claim later.

- For the use of more costly labour-only specialist trade sub-contractors.

- The type of work associated with daywork expenditure may require a higher level of supervision.

Figure 9.3 Calculation of daywork labour rates

Description	Quantity	Rate	Prime cost of labour for daywork £	Items not included in definition of prime cost
(a) Annual prime cost of labour				
Guaranteed minimum wage	1870	4.58	8564.60	
Extra payment for skill				
Public holidays	8	35.72	285.79	
Employer's national insurance contribution	8565	7%	599.55	
Contribution to annual holiday credits	47	17.55	824.85	
Contributions to death benefit scheme	47	2.05	96.35	
Contributions, levy, tax payable by employer	8565	0.25%	21.41	
Annual total for labour			10,392.55	
Divided by 1870 hours for hourly rate			5.56	
(b) Incidental costs overheads and profit				
Head office charges				☐
Site staff				☐
Trade supervision not working manually				☐
Additional cost of overtime				☐
Time lost due to inclement weather				☐
Additional bonuses and incentive schemes				☐
Subsistence and periodic allowances				☐
Fares and travelling allowances				☐
Sick pay				☐
Third party and employer's liability insurance				☐
Liability in respect of redundancy payments				☐
Tool allowances				☐
Use, repair and sharpening of non-mech tools				☐
Use of erected scaffolding, staging, trestles				☐
Use of tarpaulins, protective clothing, artificial lighting, safety and welfare, etc.				☐
Profit				☐

Calculation of percentage required for incidental costs

All-in hourly rate (see Fig. 7.1) = 7.14
  add site staff costs (7%)
  add head office charges (5%)
  add profit (5%)

All-in rate required for labour = 8.42
  less prime cost (given in table) = 5.56

Addition required = 8.42 − 5.56 = 2.86

Percentage addition required for incidental costs = **2.86/5.56 = 51%**

With daywork charges included in a competitive tender, the contractor must look for ways to minimise costs. If the provisional sum for dayworks is large then contractors may insert lower percentage additions. Conversely, where insufficient sums are allowed for daywork, contractors may add greater percentages and assess the possible over-recovery of margin in the final review meeting.

Materials and goods

The prime cost of materials and goods obtained from stockists or manufacturers is the invoice cost after deduction of all trade discounts but including cash discounts not exceeding 5 per cent and including the cost of delivery to site. Since the material cost is that which is invoiced by the supplier, it may be assumed that the prime cost includes for overlaps and waste. Similarly, the costs of packing materials, handling unloading, storage and returning packing cases are also part of the prime cost allowance.

The prime cost of materials and goods supplied from the contractor's stock is based upon the current market prices, plus any appropriate handling charges.

The percentage addition will be based on the contractor's allowance needed for site and head office overheads and profit. Typical amounts quoted are 10 per cent for site overheads, 5 per cent for head office overheads and 5 per cent for profit. For a particular project, an estimator may take the view that some of these costs will not be incurred and so can be ignored for the purpose of winning the contract.

Plant

The RICS/BEC definition includes a 'Schedule of basic plant charges'. The rates in the schedule are intended to apply solely to daywork carried out under and incidental to a building contract. They are *not* intended to apply to:

- Jobbing or any other work carried out as a main or separate contract.

- Work carried out after the date of commencement of the Defects Liability Period.

The rates apply to plant and machinery already on site, whether hired or owned by the contractor. Unless otherwise stated, the rates include the cost of fuel of every description (lubricating oils, grease), maintenance, sharpening of tools, replacement of spare parts, all consumable stores and for licences and insurances applicable to items of plant. They do not include the costs of drivers and attendants.

The rates must be applied to the time during which the plant is actually engaged in daywork. Whether or not plant is chargeable on daywork depends on the daywork agreement in use. The inclusion of an item of plant in the schedule does not necessarily indicate that the item is in fact chargeable.

Rates for plant not included in the schedule, or not on site but specifically hired for daywork, shall be settled at prices which are reasonably related to the rates in the schedule, having regard to any overall adjustment quoted by the contractor in the conditions of contract. (*Note*: Items such as small plant and loose tools, non-mechanical based tools, erected scaffolding, staging, trestles and the like are excluded from the 'Schedule of basic plant charges'. They, together with consumable stores and protective clothing, are normally incorporated into the percentage addition added to the labour rate.)

The contractor will need to allow for various additions to the basic plant rates when compiling the daywork rate for plant. These include:

- General overheads, incidental costs and profit.

- The effects of inflation since the publication date of the 'Schedule of basic plant charges'.

- The differences between rates given in the schedule and the anticipated rates actually being paid for plant.

10. PROJECT OVERHEADS

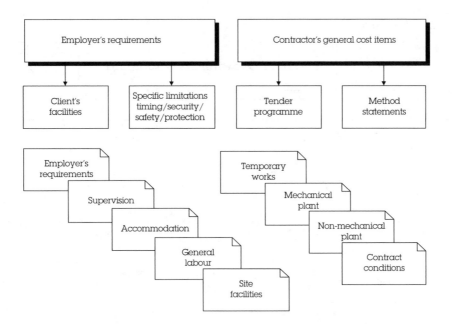

10.1 INTRODUCTION

The standard methods of measurement for civil engineering and building give the general items which should be described in a bill of quantity, in two main parts: the specific requirements of the employer and the facilities which must be provided by the contractor to carry out the work. Items listed in the preliminaries/ general conditions section of a bill of quantity are not exhaustive and are provided to produce a framework for the convenience of pricing.

The actual supervision, site facilities, plant and temporary works to be employed will be decided by the contractor unless it is clearly stated that certain systems must be used. A temporary roof may be required above an historical building, for example, if the employer is convinced that such an arrangement is the only way to protect the valuable contents of a building during a re-roofing contract. On the other hand, the decision to construct a temporary roof may be left to the tenderers.

Figure 10.1 shows a checklist used by estimators to list the items for inclusion in the project overheads. The points come from three sources:

- A careful reading of the preliminaries bill.

- Various items noted during the pricing stage.

Figure 10.1 Checklist

KEYSTONE CONSTRUCTION LTD	**PROJECT OVERHEADS CHECKLIST**	Tender number 97/201

Ref.	Description	Section	Complete
1.3d	Bond required 10% for 24 months	Contract conditions	✓
1.4c	Principal contractor – prepare safety plan	Miscellaneous	✓
1.7f	Liquidated damages £5,500 per week	Contract conditions	✓
1.12d	Office for clerk of works 15 m² plus furniture	Empl requirements	✓
1.12f	Telephone for clerk of works – install and charges	Empl requirements	✓
1.15a	Co-ordination of services – qualified engineer	Supervision	✓
1.17g	Fax machine required on site	Site facilities	✓
1.18f	Rubbish to be removed daily	Labour/plant	✓
1.18h	Hoarding required to south boundary – drg 406	Temp works	✓
1.18k	Notice board	Temp works	✓
1.20c	Protection for conc columns	Temp works	✓
1.24b	Final clean check phased completion	General labour	✓
	ATTENDANCES		
S14	Check additional scaffolding for ceilings	Non-mech plant	✓
S16	Check mixer for ceramic tiling	Mech plant	not required
	GENERALLY		
	Check concrete skip for crane	Mech plant	✓
	Check additional disposal for piling – 20 skips	Non-mech plant	✓

- The preparation of method statements and programme.

SMM7 defines a 'fixed charge' as the cost of work which is to be considered as independent of duration, and a 'time-related charge' as the cost of work which is to be considered as dependent on duration. Items in bills of quantity for employer's requirements and contractor's general cost items are given for both categories. The purpose is to improve the allocation of values in valuations for interim payments and variations. Some bills of quantity list 'value-related items' for general items such as insurances, bonds and the supply of water to the site on the bill summary page.

10.2 PROGRAMME

The tender programme is a vital tool for the calculation of project overheads. The cost of many items will vary in direct proportion to the contract period and the length of time that staff, facilities, plant and temporary works will be required on the site.

It is normal for costs to be built up on a fixed-charge and a time-related basis. For example, in setting up a tower crane a high fixed charge is met at the start, there follows a recurring cost related to the time the equipment is on site. There will then be a subsequent fixed cost for removing the equipment on completion. By estimating such items in these two distinct forms, changes to the programme can be evaluated quickly, by reference to the time-related costs. The estimator will be able use the tender programme for most of his calculations or develop a project overheads programme, as shown in the example in Figure 10.2.

A carefully prepared tender programme will provide a contractor with the opportunity to reduce the contract duration leading to cost savings for staff, temporary works and plant. Since prices for the measured portion of a bill will be similar for all contractors, it is this innovative planning skill which produces a competitive advantage over other tenderers. If, on the other hand, the client wishes to retain a longer contract period (to avoid claims for extension of time, for example) the contractor can be at risk if he allows sufficient staff for only part of the time.

10.3 PROJECT OVERHEADS SCHEDULE

The estimator's notes must now be brought forward for consideration and entered in the various sections of the 'Projects overheads summary' (see Figure 10.3). The various parts of this schedule are shown in Figures 10.4 to 10.13. In addition, other items which have not been included in unit rates must also be considered and costs established.

The items shown in the example forms should not be considered as absolute. Items may be added to or omitted from those listed according to the preferences and policies of individual companies. The forms provided illustrate typical items which are required in project overheads. Various techniques exist to price these items and in some organisations the project overheads are worked back into unit rates. This is a matter for each company to decide. The total estimate of the cost of project overheads is transferred from this schedule to the final summary document.

PROJECT OVERHEADS PROGRAMME

KEYSTONE CONSTRUCTION LTD

Project
Helix Laboratories, Westfield

Description	Month Date Week nr	-4	-3	-2	-1	1	2	3	4	5	6	7	8	9	10	11	12	13	14	15	16	17	18	19	20	21	22	23	24	25	26	27	28	29	30	31	32	33	34	35	36	37	38	39	40	41
Mobilisation																																														
Award																																														
Off-site mobilisation																																														
Set up site																																														
Sub-structure																																														
Piling																																														
Foundations																																														
Drainage																																														
Slab																																														
Superstructure																																														
Steelwork																																														
Upper floors																																														
Roofing																																														
External walls																																														
Internal fabric																																														
Internal fittings																																														
Services																																														
Finishes																																														
External works																																														
Site works																																														
Drainage																																														
Clear site																																														

Project overheads	wks																																													
Employer's requirements	40																																													
Site manager	40																																													
Foreman	38																																													
Engineer	26																																													
Planning engineer	4																																													
Quantity surveyor	43																																													
Clerk/typist	36																																													
Accommodation/Site facilities	40																																													
Tower crane	11																																													
Scaffolding	12																																													
Hoists	12																																													
Forklift	16																																													
Dumper	31																																													

Figure 10.2 Programme

KEYSTONE CONSTRUCTION LTD	PROJECT OVERHEADS SUMMARY												Project — Helix Laboratories, Westfield
	Time-related costs						Fixed costs						Totals
	Lab	Pit	Mat	Sub	O'heads	Totals	Lab	Pit	Mat	Sub	O'heads	Total	
1 Employer's requirements		3900			600	4500	600	700	400	200	200	2100	6600
2 Supervision					108,500	108,500							108,500
3 Accommodation		9430				9430	900	1940				2840	12,270
4 General labour	18,500					18,500	10,300					10,300	28,800
5 Site facilities					6170	6170		600	1600		400	2600	8770
6 Temporary works	1500	1400	3175	100		6175							6175
7 Mechanical plant		18,420				18,420		3000				3000	21,420
8 Non-mechanical plant		18,020				18,020		1900				1900	19,920
9 Contract conditions											1000	1000	1000
10 Miscellaneous					1000	1000	90		490		1240	1820	2820
Totals to Final Summary	20,000	51,170	3175	100	116,270	190,715	11,890	8140	2490	200	2840	25,560	216,275

Figure 10.3 Projects overheads summary – Form E

EMPLOYER'S REQUIREMENTS

KEYSTONE CONSTRUCTION LTD

Project — Helix Laboratories, Westfield

Ref.	Type	Grade	Programme From	Programme To	Quant	Wks	Rate	Total Lab	Plt	Mat	Sub	O'heads	£
		(Lighting and heating in Site Facilities)											
	Time-related costs												
	Accommodation	Offices	1	40	2	80	30		2400				2400
		Conference room											
		Toilets	1	40	1	40	20		800				800
		Stores											
		Laboratories											
	Equipment	Furniture	1	40	1	40	5		200				200
		Telephone rental	1	40	1	40	5					200	200
		Telephone calls	1	40	1	40	10					400	400
		Testing equipment											
		Surveying instruments											
	Charges	Rates											
	Attendance	Surveying assistance											
		Drivers											
		Lab assistants											
		Cleaners	1	40	0.25	10	50		500				500
		Total time-related costs							3900			600	4500
	Fixed costs												
	Mobilisation	Transport and cranage	1				250		250				250
		Foundations and drainage	1					200		200			400
		Erection and fitting out	1					200		200	200		600
	Equipment	Telephone installation									200		200
		Facsimile installation							200				200
	Charges	Land acquisition											
	Demobilisation	Dismantling	40					200					200
		Cranage and transport	40						250				250
		Total fixed costs						600	700	400	200	200	2100

Figure 10.4 Employer's requirements

KEYSTONE CONSTRUCTION LTD — **SUPERVISION** — Project: *Helix Laboratories, Westfield*

Ref.	Type	Grade	Programme From	Programme To	Nr.	Wks	Total wks	£/wk	Total costs £ Time related	Total costs £ Fixed
	Management	Project manager								
		Site manager	−2	38	1	40	40	750	30,000	
	Supervision	General foreman								
		Foreman	3	40	1	38	38	650	24,700	
		Trades foreman								
		Ganger								
	Engineering	Senior engineer								
		Engineer	1	20	1	20	20	575	11,500	
		Assistant engineer	28	33		6	6	400	2400	
	Planning & co-ordination	Planning engineer	−2	2	1	4	4	650	2600	
		Co-ordination engineer								
		Services engineer								
	Quantity surveying	Senior quantity surveyor	−2	41	1	43	43	700	30,100	
		Quantity surveyor								
		Assistant quantity surveyor								
	Support staff	Secretary								
		Clerk/typist	2	38	1	36	36	200	7200	
		Storeman/checker								
		Safety officer								
		Security/watchman								
		Canteen staff								
		First aider								
	Others									
		Totals					187		108,500	

Figure 10.5 Supervision

KEYSTONE CONSTRUCTION LTD		ACCOMMODATION						Project: Helix Laboratories, Westfield						
Ref.	Type	Grade	Programme From	To	Quant.	Wks	Rate	Total Lab	Plt	Mat	Sub	O'heads	£
	Time-related costs												
	Office accommodation	Management	1	38	1	38	35		1330				1330
		Supervision	1	40	1	40	25		1000				1000
		Engineering	1	33	1	33	30		990				990
		Planning/Co-ordination											
		Quantity surveying	1	40	1	40	30		1200				1200
		Support staff											
	Ancillary accommodation	Stores	1	40	2	80	25		2000				2000
		Canteen/welfare	1	38	1	38	45		1710				1710
		Toilets	1	40	1	40	30		1200				1200
		Drying room											
		First aid											
		Workshops											
	Charges	Rates											
		Total time-related costs							9430				9430
	Fixed costs												
	Mobilisation	Transport to site			8		40		320				320
		Cranage			2		250		500				500
		Foundations			1								
		Drainage			1								
		Erection			1		300	300					300
		Fitting out and decoration			1		300	300					300
		Furniture and fittings			1		300		300				300
	Charges	Land acquisition											
	Demobilisation	Dismantling			1		300	300					300
		Cranage			2		250		500				500
		Transport from site			8		40		320				320
		Total fixed costs						900	1940				2840

Figure 10.6 Accommodation

KEYSTONE CONSTRUCTION LTD — **GENERAL LABOUR** — Project: *Helix Laboratories, Westfield*

Ref.	Type	Grade	Programme From	Programme To	Nr.	Wks	Total wks	£/wk	Total costs £ Time related	Total costs £ Fixed
	(Items not included in rates)									
	Attendant labour	Unloading	16	32	1	17	17	300	5100	
		Distribution								
		Setting-out assistants	1	13	1	13	13	200	2600	
	Plant labour	Drivers								
		Fitters								
		Pump maintenance								
		Banksmen								
		Tower crane banksmen								
		Hoist operator	17	28	2	12	24	300	7200	
	Temporary works	Scaffold erection	14	16	3	3	9	300		2700
		Scaffold adaptation	17	28	1	12	12	300	3600	
		Scaffold removal	29	30	3	2	6	300		1800
		Fencing adaptation								
		Shoring adaptation								
		Electrician								
	Protection	Install			1	4	4	300		1200
		Take away			1	2	2	300		600
		Repair								
	Cleaning	Daily								
		Progressive								
		Final	39	40	2	2	4	300	1200	
	Snagging	Carpenter								
		Painter	39	40	1	2	4	350		1400
	General attendance		39	40	1	2	4	350		1400
	Others									
		Totals					99		18,500	10,300

Figure 10.7 General labour

KEYSTONE CONSTRUCTION LTD — **SITE FACILITIES** — **Project** Helix Laboratories, Westfield

Ref.	Type	Grade	Programme From	Programme To	Quant.	Wks	Rate	Total Lab	Plt	Mat	Sub	O'heads	£
	Time-related costs												
	Administration	Stationery and postage				40	20					800	800
		Office equipment				40	30					1200	1200
		Computers				40	10					400	400
		Humidity and temperature control											
	Services – rental and charges	Telephone				40	30					1200	1200
		Accommodation power and lighting				40	20					800	800
		Heating				40	20					800	800
		Water				40	10					400	400
		Task power and lighting	22	40		19	30					570	570
		Security lighting											
		Total time-related costs										6170	6170
	Fixed costs												
	Mobilisation, installation, removal and reinstatement	Office equipment								200			200
		Computers								1000			1000
		Humidity and temperature control											
		Telephone										400	400
		Accommodation power and lighting											
		Heating								200			200
		Water – supply											
		Water – standpipes and tanks							200				200
		Task power and lighting							400				400
		Security lighting											
		Temporary drainage								200			200
		Fuel storage											
		Total fixed costs							600	1600		400	2600

Figure 10.8 Site facilities

KEYSTONE CONSTRUCTION LTD			**TEMPORARY WORKS**						Project		Helix Laboratories, Westfield		

Ref.	Type	Grade	Programme		Quant.	Wks	Rate	Total					£
			From	To				Lab	Plt	Mat	Sub	O'heads	
	(For scaffolding see General labour and Non-mech plant)												
	Time-related costs												
	Maintenance and adaptation	Access roads											
		Hardstandings											
		Compounds											
		Traffic control											
		Hoardings											
		Temporary fencing											
		De-watering											
		Temporary structures											
		Safety barriers											
		Total time-related costs											
	Fixed costs												
	Mobilisation, installation,	Access roads			240	m²	10	600	600	1200			2400
	removal and reinstatement	Hardstandings			150	m²	10	400	400	700			1500
		Compounds											
		Traffic control											
		Hoardings and temporary fencing			85	m	15	400		875			1275
		Safety barriers											
		Construction joints							400				400
		De-watering											
		Temporary structures											
		Shoring and centring											
		Notice boards			1	nr	300	100		100	100		300
	Protection	Equipment and materials								300			300
	Services	Diversions											
		Protection											
		Total fixed costs						1500	1400	3175	100		6175

Figure 10.9 Temporary works

KEYSTONE CONSTRUCTION LTD		MECHANICAL PLANT							Project — Helix Laboratories, Westfield		
			Programme							Total costs £	
Ref.	Type	Grade	From	To	Quant.	Wks	Total wks	Rate	Time related	Fixed
	Lifting plant	Tower crane and driver								
		Mobile crane and driver	12	16	1	5	5	1300	6500	
		Goods hoist	17	28	2	12	24	100	2400	
		Passenger hoist								
	Transport	Forklift	16	31	1	16	16	200	3200	
		Dumper	8	38	1	31	31	100	3100	
		Tractor and trailer								
		Van/car								
	Concreting plant	Concrete pump								
		Compressor and tools	5	20	1	16	16	100	1600	
		Powerfloats, tamping equipment	9	20	1	12	12	50	600	
	Mixers	Concrete								
		Mortar	8	38	1	31	31	20	620	
	Plant reconciliation	Additional excavator time								
		Additional lorries time								
		Additional rollers time								
		Additional water pumping	5	12	1	8	8	50	400	
	Setting up	Tower crane								
		Hoist			2			500		1000
		Transport to site			5			50		250
		Electrical connections			1			500		500
	Removal	Tower crane								
		Hoist			2			500		1000
		Transport from site			5			50		250
	For banksmen see General labour plant rates include fuel							Totals	18,420	3000

Figure 10.10 Mechanical plant

KEYSTONE CONSTRUCTION LTD — **NON-MECHANICAL PLANT** — Project: Helix Laboratories, Westfield

Ref.	Type	Grade	Programme From	Programme To	Quant.	Wks	Total wks	Rate	Total costs £ Time related	Total costs £ Fixed
	Scaffold hire	External	17	28	1200 m²		12	0.5	7200	
		Internal								
		Hoist towers			2	12	24	50	1200	
		Mobile towers			2	20	40	20	800	
		Temporary roofs								
		Debris netting/sheeting			1200 m²			3	3600	
		Roof edge protection								
		Ladders								
		Rubbish chutes			2	12	24	55	1320	
	Instruments	Level			1	26	26	20	520	
		Theodolite			1	26	26	30	780	
	Traffic control	Signs								
		Traffic lights								
		Cones								
	Miscellaneous	Waste skips			1	26	26	100	2600	
		Lifting slings and equipment								
		Additional trench supports								
		Bar benders								
		Fuel tanks								
	Transport	To site			3			50		150
		From site			3			50		150
	Small tools and equipment	(if not included in labour rates)			1	40	40	40		1600
	For scaffold erection, adaptation and removal see General labour									
								Totals	18,020	1900

Figure 10.11 Non-mechanical plant

KEYSTONE CONSTRUCTION LTD		CONTRACT CONDITIONS						Project Helix Laboratories, Westfield			
										Total costs £	
Ref.	Type	Grade	Programme		Total wks	£/wk	Base cost £	%	Time related	Fixed
			From	To						
	(For fluctuations firm price adjustment form)									
	Insurances	All-risks insurance					2,300,000	0.5	see final summary	
		Employer's liability					110,000	0.2	see final summary	
		Public liability					110,000	0.2	see final summary	
		Excesses								1000
	Bonds	Performance					230,000	1.5	see final summary	
		Parent company								
	Warranties	Contractor								
		Sub-contractor								
		Designer								
	Special conditions									
	Professional fees	Architect								
		Engineer								
		Quantity surveyor								
		Land surveyor								
		Legal								
		Geologist								
		Progress photographs								
	Statutory fees	Building regulations								
		Planning consent								
		Permits and parking charges								
	Others									
								Totals		1000

Figure 10.12 Contract conditions

KEYSTONE CONSTRUCTION LTD			MISCELLANEOUS					Project — Helix Laboratories, Westfield					
Ref.	Type	Grade	Programme		Quant.	Wks	Rate	Total					£
			From	To				Lab	Plt	Mat	Sub	O'heads	
	Time-related costs												
	Winter working	Weather protection											
		Additional lighting											
		Reduced productivity											
		Frost control											
		Protective clothing											
	Quality assurance	Administration											
	Safety	Documentation/safety file										1000	1000
		Training											
		Total time-related costs										1000	1000
	Fixed costs												
	Setting out consumables	Pegs and profile boards								200			200
		Concrete and steel markers								200			200
	Testing and samples	Concrete testing			55		8					440	440
		Other testing											
		Sample panels			3		60	90		90			180
		Material samples											
	Safety	Safety plan											
		Hard hats and clothing			20		40					800	800
		Total fixed costs						90		490		1240	1820

Figure 10.13 Miscellaneous

Contractors and firms that carry out small repetitive works often apply a fixed percentage for project overheads. A figure is found from experience to give sufficient income to pay for overheads. The pitfall with this approach is that site overheads, particularly supervision, will be proportionately higher for small-value jobs which are to be carried out to an extended programme. It is also necessary to price temporary works such as scaffolding separately because they are rarely related to the value of the works.

Various groupings of items relating to project overheads are possible according to company preference. The sample forms consider establishing costs on a multi-column basis in order to allow the build-up of costs in a similar way to unit rates in terms of labour, plant, materials and sub-contractors. Not all columns are relevant to all sections but this consistent approach and layout allows costs to be established under these major headings. The following are the main groups of items to be considered (refer to Figure 10.3).

1 Employer's requirements (Figure 10.4)

This category of overheads is used for the accommodation, equipment and services needed by the employer's representative when based on site. SMM7 states that heating, lighting, cleaning and maintenance of employer's accommodation are deemed to be included. The term 'employer's requirements' is used in SMM7 to include other items such as quality standards, safety, security, protection, limitations on method and programme, specific temporary works and services. These costs are entered on the schedule.

2 Supervision (Figure 10.5)

Since site management and staff form the largest part of project overheads, the estimator must propose the smallest team with the skills needed to produce a successful project. This will include all personnel whose costs have not been included in unit rates or head office overheads, whatever their method of remuneration. Care must be taken to provide supervision of nominated sub-contractors and work described as 'defined' provisional sums.

The costs allowed will normally include all costs associated with employing the staff, such as company cars and running costs, pension scheme contributions, subsistence allowances, bonuses and so on. Monthly all-in staff costs are normally provided by management as average rates in a form which disguises actual rates of pay for individual members of staff.

If visiting head office staff (such as quantity surveyors, contracts managers, safety managers and quality advisers) are charged for projects as part of the company's policy, the estimator must liaise with management over the extent of such head office services to ensure that due allowance is made in the cost estimate.

3 Accommodation (Figure 10.6)

The hire or purchase of accommodation should be considered under this section. Reference to the method statement and programme will be necessary to determine the intended layout of the site and time period for the various elements of accommodation. Transportation, erection, fitting out and decoration, subsequent dismantling and reinstatement is noted in the fixed cost section of the form. Repetitive items such as rental charges and local authority rates are noted under time-related costs.

The connection of main services and charges are noted under item 5, 'Site facilities'.

It is often difficult to predict when accommodation is taken away or moved to another part of a site. A site office and storage container may be needed after practical completion to complete minor works and snagging.

Site buildings are obtained from three sources:

- Hired accommodation can be obtained locally.

- Site huts can be purchased especially for long-term projects.

- Temporary buildings may be available from the contractor's own stock.

If alterations are proposed to an existing building, using space inside the building can provide a saving, even after an allowance is made for fitting out and finishes. In some cases offices in adjacent buildings can be rented which brings the benefit of having more space on site for building operations. Whichever type of office is used, some fitting-out costs are needed, and on long-term projects maintenance costs may be incurred.

An allowance is needed for local authority charges whichever arrangement is used. These are for rates and the use of land beyond the boundary of the site.

4 General labour (Figure 10.7)

General labour can be viewed as a common resource to service all trades and sub-contractors on a site. This can include keeping the site tidy, unloading materials, distributing materials and driving mechanical plant. The proportion of attendant labour in unit rates and that priced in project overheads must be assessed at the review meeting to reconcile the needs of direct labour, labour-only sub-contractors and trade specialists. As a guide, it is better to price attendant labour on a time-related basis and include only direct fixing costs in unit rates.

Other items such as cleaning the building during the works and final cleaning on completion are also normally priced in project overheads. Attendant labour on sub-contractors can also be priced in project overheads if such items are not included in unit rates. The drivers of general plant can also be priced here if not included in the price of the plant item.

5 Site facilities (Figure 10.8)

It is sometimes difficult to distinguish between site facilities and temporary works. In these schedules site facilities include the installation and removal of services and associated consumable items. Other facilities such as temporary structures and hardstandings, which are directly associated with the nature of the works, are entered in the temporary works sheet.

Installation charges are obtained from statutory undertakings and resource usage can be calculated using outputs and current rates. Alternatively the cost of site facilities is valued in relation to the number and type of staff employed on the site. For example, it might be found from historical records that a site manager needs an allowance of £30 per week for telephone calls and an annual computer allowance of £500. A further approximation would be to establish a cost for site facilities as a percentage of supervision. In the schedules it can be seen that site facilities amount to 8 per cent of staff costs.

6 Temporary works (Figure 10.9)

The costs of temporary access roads, hoardings and temporary structures are determined in this section. Recurring costs to maintain access roads and move fencing, for example, should be calculated and added to the costs of final removal and reinstatement of the site. The method statement and tender programme are important in establishing the systems and durations for these items (see Chapter 6, page 69).

The route of temporary access roads and hardstandings needs careful planning so that use may be made of existing roads and permanent materials. With this in mind, an allowance is needed for reinstatement or making up levels when the work is completed.

In this Code, scaffolding has been included under the heading 'Non-mechanical plant'. Large scaffolding systems including temporary roofs, falsework and crash decks are provided by specialist sub-contractors, and often priced on the temporary works schedule.

Materials for protecting finished work must be included. This might be for the protection of early structural components such as fair faced columns, or whole areas of finished work which can suffer damage during the commissioning stage of a contract.

7 Mechanical plant (Figure 10.10)

Mechanical plant is included in the project overheads for two reasons:

- Plant which is common to more than one trade can be viewed as common plant available to all. On a small site, for example, one mortar mixer can service the bricklayers, drain-layers and plasterers laying screeds. In fact one mixer may be needed for a large part of a contract duration.

- To reconcile the plant included in measured rates so that plant resources can be rationalised. A backacter, for example, would not be used for a few days, taken off site for a day and then brought back for intermittent periods. By using the same machine for drainage and earthworks it may be possible to usefully employ one machine for an extended period.

The method statement and programme will give valuable guidance at this stage and indicate the type of plant required and the time the plant will be required on site.

Specific temporary works associated with plant must also be established with allowances made if not already covered under section 6, 'Temporary works', above.

A special electricity supply may be needed for large plant such as a tower crane and the estimator must bear in mind the installation, running costs and possible extra electricity charges.

The estimator must produce all-in rates for many of these items of plant. Drivers or tradesmen, where needed, can either be incorporated as part of the all-in rate or, alternatively, be placed in section 4, 'General labour'. Attendant labour for mechanical plant, such as operating pumps and hoisting equipment together with down time for servicing, should be established by the estimator.

If prices are required from plant hirers, enquiries for general plant must be sent out giving the type of plant required, the capacity, programme and in the case of cranes the location and layout of the site.

8 Non-mechanical plant (Figure 10.11)

The largest item of non-mechanical plant is scaffolding and ancillary equipment, such as debris netting and rubbish chutes. Alternatively, scaffolding may be treated in the following ways:

- As a domestic sub-contract on the sub-contract summary form.

- In the sub-contract column on the temporary works schedule.

- Part of the build-up of a unit rate.

Small (hand) tools and equipment may be included in the all-in rate for labour since their value must be in proportion to the amount of labour employed on the works; *but are not included in the all-in rate example in Chapter 7.*

9 Contract conditions (Figure 10.12)

The preliminaries section of the bills of quantity and the actual conditions of contract must be scrutinised to ensure that all items attracting a monetary value have been included in the cost estimate.

The cost of bonds, insurances and professional services (where applicable) should be recalculated after the final review meeting as a percentage of the contract value. For large projects, or those in hazardous environments or near watercourses, a quotation for insurance premiums is needed. Not all losses and damage on a site can be recovered under an insurance policy. For losses which have a value below the agreed insurance excess, an allowance should be added to the tender, based on historical company records for unrecovered losses.

Allowances for firm or fluctuating price are considered separately when the production costs and project overheads are known. Other matters concerning the contract conditions will also be considered under this heading.

10   Miscellaneous (Figure 10.13)

Any unusual features associated with the project which do not fall naturally into the defined categories can be priced in this section.

Safety and quality assurance procedures will be an everyday part of a site manager's work, and should not bring about additional costs on small and medium-sized contracts. A great deal of additional documentation will be required, however, under the CDM regulations; in particular the development of the pre-tender health and safety plan throughout the construction phase and the preparation of data for the safety file by the end of a large project.

Although quality assurance procedures and safety systems are well established, there are additional overhead costs which are included in the head office overheads. These are for specialist managers and consultants used to support in-house staff. It may be, with particularly complex schemes, that these costs are directly apportioned to specific projects.

When unit rates for labour were being examined in Chapter 7 it was noted that certain costs attributable to the employment of labour were often contained in the project overheads. Some costs relating to labour are not clear at the early stages of preparation of the estimate when the volume of labour required on the project has not been accurately established. These items, which have not been included in the unit rate calculation, must now be brought forward for pricing in the project overheads.

The allowance made for travel and fares, subsistence allowance, attraction money and exceptional inclement weather will be based upon information gained from the site visit when an appreciation of the labour availability will have been made. This item, together with attraction money and additional bonus allowances, are critical areas for contractors and the estimator must take considerable care in making recommendations to management in such matters.

11. COMPLETING COST ESTIMATES

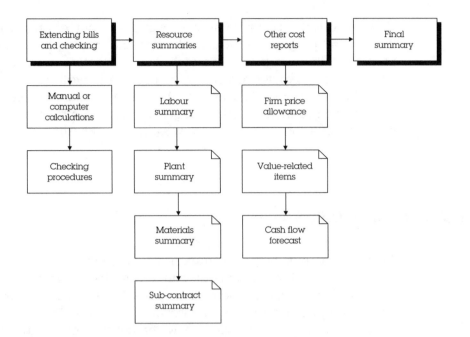

11.1 EXTENDING BILLS OF QUANTITY

When all the net unit rates have been completed, the bills must be extended and totalled before project overheads and other allowances are calculated. This gives the estimator an overall picture of the main elements of the work. Such extensions are made with separate sub-totals being produced for the four basic elements of labour, plant, materials and domestic sub-contractors.

The extension of the bill rates to totals, collections and summaries must be carried out in a manner to eliminate clerical error and also to allow the establishment of various sub-totals relating to elements of the work or trades as necessary.

With computer-estimating systems, reconciliation of the various printed reports will now be possible. The total of the bills of quantity can be gained from different reports, in the following ways:

- The analysis of the bill total into totals for labour, materials, plant, sub-contractors (and PC and provisional sums if they have been entered into the computer).

- A total for each of the resources used in building-up the estimate.

- Totals for each element of a building if an elemental bill is used.

These reports should be compared with the resources envisaged on the tender programme to check that the assumptions used in preparing the programme match the estimator's resources. For example, a computer report might give the overall quantity for a 20 t backacter as 74.2 hrs. If the tender planner, after an examination of methods, has suggested that the machine needs to be available on site for two weeks, an adjustment must be considered.

11.2 NET COST SUMMARIES

An estimator must prepare summaries of the resources making up an estimate so that management can assess the sources of labour, plant, materials and services, their price levels, price comparisons between suppliers and possible discounts available. The forms given in Figures 11.1 to 11.4 illustrate how the constituents of an estimate are summarised clearly in a way which helps management make decisions quickly using accurate information.

These forms can be hand-written or produced by a computer-aided estimating system; whichever is adopted, when the forms are complete the computer information should not be changed. Hand-written forms are being superseded by tailor-made computer reports but are retained by organisations which want their estimators to thoroughly check the information from computer printouts.

Adjustments (see Figure 11.5) must be made on the forms or added to a final review adjustment form (see Figure 11.6). A record must be kept showing the changes made during the final review stage.

The final summary brings together all the parts of an estimate, with costs apportioned between labour, plant, materials, sub-contractors, site overheads and PC and provisional sums. The summary given in Figure 11.7 shows the entries as net costs, the estimator has taken out all the discounts offered by suppliers and sub-contractors.

11.3 PRICE FLUCTUATIONS

In times of relatively low and predictable inflation, contractors are expected to submit tender prices which remain fixed for the anticipated duration of the work.

LABOUR SUMMARY

KEYSTONE CONSTRUCTION LTD

Project
Helix Laboratories, Westfield

Ref.	Description	Trade	Total hrs	All-in rate £/hr	Totals in rates £	Typical outputs	Labour-only			Late quotations		
							Name	Quote	Adjustment	Name	Quote	Adjustment	Adjusted totals
L01	Concrete work	Conc labourer	1200	6.00	7200	fnds 2.0 hr/m³, slab 1.6 hr/m³	MK Formwork	6900	−300	F. Smith	6653	−247	6653
L02		Carpenter	300	9.00	2700	fnds 0.5 hr/m³, slab 0.4 hr/m³	ditto	inc	−2700	ditto	inc		0
L03	Brickwork	Bricklayer	1100	9.00	9900	facings £250/th inc labourer							9900
L04		Br labourer	550	6.00	3300								3300
Totals			3150		23,100				−3000			−247	19,853

Figure 11.1 Labour summary – Form A

PLANT SUMMARY

Project
Helix Laboratories, Westfield

KEYSTONE CONSTRUCTION LTD

Ref.	Resource type	Quant.	Unit	Output /hr	unit	Duration hrs	Rate £/hr	Totals in rates £	Name	Quote	Adjustment	Name	Quote	Adjustment	Adjusted totals
										Sub-contract			Late quotations	
P01	JCB3, Founds	5800	m³	6	m³	967	14.00	13,533	} Thomas	28,500	−1833	F. Smith	26,350	−2150	26,350
P02	Roller, founds	7200	m²	10	m²	720	5.00	3600	} ditto	inc		ditto	inc		
P03	JCB3, hardcore	6000	m³	8	m³	750	14.00	10,500	} ditto	inc		ditto	inc		
P04	Roller, hardcore	10,800	m²	20	m²	540	5.00	2700	} ditto	inc		ditto	inc		
P05	20t tipper	5800	m³	8	m³	725	22.00	15,950	} Thomas	17,850	−4142	F. Smith	16,950	−900	16,950
P06	Dozer on tip	5800	m³	24	m³	242	25.00	6042	} ditto	inc		ditto	inc		
Totals						3943		52,325			−5975			−3050	43,300

Figure 11.2 Plant summary – Form B

KEYSTONE CONSTRUCTION LTD		**MATERIALS SUMMARY**								Project — Helix Laboratories, Westfield				
											Late quotations			
Ref.	Resource type	Supplier	Quant. BOQ	Unit	Convert factor t/m³	Waste %	Total quant	Net rates £	Totals in rates £	Name	Quote	Adjustment	Adjusted totals
M01	Hardcore	Target Stone	6000	m³	1.90	10.00	12,540	7.00	87,780				87,780
M02	Type 1 subbase	Target Stone	7200	m³	2.10	10.00	16,632	6.75	112,266				112,266
M03	Concrete 20N	MBC Concrete	250	m³		7.00	268	45.25	12,104				12,104
M04	Concrete 30N	MBC Concrete	300	m³		5.00	315	52.60	16,569				16,569
M05	Concrete sundries	Various	1	sum					4865				4865
M06	Formwork	Various	1	sum					1850	F. Smith	1500	–350	1500
M07	Facing bricks	Dexter Brick	55.4	th		6.00	59	335.00	19,673		inc		19,673
M08	Blocks 140 mm	Opus Supplies	6550	m²		7.50	7041	9.65	67,948				67,948
M09	Brick sundries	Various	1	sum					3860				3860
M10	Timber doors	Rosewood Joinery	65	nr		2.50			23,114				23,114
M11	Ironmongery	Lasermax	1	sum					11,480	Opus Supplies	11,020	–460	11,020
								Totals	361,509			–810	360,699

Figure 11.3 Materials summary – Form C

KEYSTONE CONSTRUCTION LTD | **SUB-CONTRACT SUMMARY** | **Project** Helix Laboratories, Westfield

Ref.	Trade	Company	Quote	Discount	Net	Attendances Lab	Plt	Mat	Inflation to form G	Late quotations Name	Net quote	Adjustment	Adjusted totals
S01	Piling	BBS	40,079	2.5	39,077					Robinson	37,080	–1997	41,074
S02	Metalwork	No quote	69,610		69,610								69,610
S03	Structural steelwork	Steelcare	105,675	2.5	103,033					PCS	99,520	–3513	106,546
S04	Roof tiling	Richard Roofing	62,629	2.5	61,063				1500				61,063
S05	Leadwork	Richard Roofing	33,570	2.5	32,731				800				32,731
S06	Metal windows	Microtex	33,001	2.5	32,176								32,176
S07	Plastering	Addford	125,141		125,141								125,141
S08	Floor coverings	Floordex	29,844	2.5	29,098				700				29,098
S09	Painting	Clover	72,225	2.5	70,419				1700				70,419
S10	Plumbing	Sharpe & Baker	96,570	2.5	94,156								94,156
S11	Mechanical	Evanbridge	255,246		255,246	500	500	400	6600				255,246
S12	Electrical	Evanbridge	279,482		279,482	500	500	400	6700				279,482
S13	Lift installation	Expert Lifts	66,103	2.5	64,450	500		400		Clearframe	61,220	–3230	67,681
Totals			1,269,176		1,255,683	1500	1000	1200	18,000			–8740	1,264,424

Figure 11.4 Sub-contract summary – Form D

KEYSTONE CONSTRUCTION LTD	**FIRM PRICE ADJUSTMENT**					Project – Helix Laboratories, Westfield					
								Totals			
Ref.	Type	Remarks	Totals	Proportion	% addition	Lab	Plt	Mat	Sub	O'heads	£
	Measured works										
	Labour		19,853		nil						
	Plant		43,300		nil						
	Materials	generally add 3% for 73% of value	292,700	73/100	3			6400			13,200
		blocks add 10%	68,000	1	10			6800			
	Sub-contractors	from sub-contract summary	1,264,424						18,000		18,000
	Attendances		4900		nil						
	Project overheads										
	Project overheads	generally 2% for 50% value	123,165	50/100	2	nil	nil	nil		300	300
	Employer's requirements										
	Supervision	add 3% for last 20 weeks	108,500	20/38	3					1700	1700
	Accommodation										
	General labour										
	Site facilities										
	Temporary works										
	Mechanical plant										
	Non-mechanical plant										
	Contract conditions										
	Miscellaneous										
								13,200	18,000	2000	33,200

Figure 11.5 Firm price adjustment – Form G

KEYSTONE CONSTRUCTION LTD	FINAL REVIEW ADJUSTMENTS					Project — Helix Laboratories, Westfield				
					Adjustments					
Ref.	Type	Remarks	From	To	Lab	Plt	Mat	Sub	O'heads	£
	Measured works									
	Labour	Reduce bricklaying	9900	8450	-1450					-1450
	Plant	Additional tip charges				6200				6200
	Materials	Reduce sub-base by £1.00/t					-16,600			-16,600
		Reduce blocks by £1.00/m²					-6550			-6550
		Adjust inflation	13,200	7900			-5350			-5350
	Sub-contractors	Generally 1% improvement						-12,600		-12,600
		Reduce inflation	18,000	8000				-10,000		-10,000
	Attendances									
	Project overheads									
	Employer's requirements									
	Supervision	Take out engineer							-11,500	-11,500
	Accommodation									
	General labour	Reduce hoist operator	7200	6350	-850					-850
	Site facilities									
	Temporary works									
	Mechanical plant	Reduce mobile crane	6500	3500		-3000				-3000
	Non-mechanical plant	Debris netting from stock	3600	2600		-1000				-1000
	Contract conditions									
	Miscellaneous	Add for winter working				1000				1000
					-2300	3200	-28,500	-22,600	-11,500	-61,700

Figure 11.6 Final review adjustments – Form H

KEYSTONE CONSTRUCTION LTD		FINAL SUMMARY					Project	Helix Laboratories, Westfield	

	Form	Net costs				Site o'heads	PC & prov sums	Sub-totals	Totals
		Lab	Plt	Mat	Sub				
Own work									
Direct work	A–C	19,853	43,300	360,699				423,852	
Domestic subs and attendances	D	1500	1000	1200	1,264,424			1,268,124	1,691,976
Project overheads									
Time related	E	20,000	51,170	3175	100	116,270		190,715	
Fixed costs	E	11,890	8140	2490	200	2840		25,560	216,275
PC and provisional sums									
Nominated suppliers	F						18,525	18,525	
Nominated subs and attendances	F	600	600	1600			139,425	142,225	
Statutory undertakers							0	0	160,750
Sub-total (estimate frozen)		53,843	104,210	369,164	1,264,724	119,110	157,950	2,069,001	2,069,001
Adjustments									
Firm price	G		13,200		18,000	2000		33,200	
Final review	H	–2300	3200	–28,500	–22,600	–11,500		–61,700	–28,500
Sub-total		51,543	107,410	353,864	1,260,124	109,610	157,950	2,040,501	2,040,501
Value-related items									
All-risks insurance, Empl & Pub Lia						11,940		11,940	
Water for the works					see proj o/h			0	
Performance bond						3450		3450	
Professional fees									15,390
Total net costs		51,543	107,410	353,864	1,260,124	125,000	157,950	2,055,891	2,055,891
Percentage breakdown	%	3	5	17	61	6	8	100	

Mark-up	%	£	
Scope		–25,000	
Risk	3	60,927	
Head office overheads	5	104,591	
Profit	5	109,820	
Sub-total		250,338	2,306,229
Provisional sums from form J			96,000
Dayworks from form J			60,800
Tender total excl VAT £			2,463,029

Signatures

Figure 11.7 Final summary

Fluctuations in cost

The method of calculation of fluctuations in costs is set out in the appendix to the conditions of contract. Fluctuations in cost can be 'full', 'limited' or none at all. It should be borne in mind that fluctuations in cost give rise to both increases and decreases and careful consideration is needed to reconcile such changes to the contractor's cost estimate.

Full fluctuations

Full fluctuations in cost are intended to provide an equitable means of reflecting changes in cost throughout the duration of a contract. Although a number of full fluctuations clauses are available in standard forms of contract, they are rarely used today. Where the formula method of fluctuations recovery is proposed, the estimator needs to understand the rules and must ensure that any shortfall or non-recoverable element is identified and also any potential for over-recovery highlighted. Due allowance should be made on the firm price adjustment form (Figure 11.5) or resource schedules for any changes needed. Adjustments using the formula method are limited to national increases and do not, therefore, reflect local market forces.

Limited fluctuations (statutory items)

This is normally limited to specific changes to rates of contribution, levies and taxes in the employment of labour and in the rates and duties and taxes on the procurement of materials. The extent of fluctuations of such costs are limited and are effectively those which arise under and by virtue of an Act of Parliament.

No fluctuations

Not an option in most standard forms of contract because there is no sensible reason why contractors should add a contingency for a new tax or statutory levy which cannot be predicted at tender stage. Nevertheless, it is not uncommon for all fluctuations clauses to be deleted and a statement added that the price is to be firm in all circumstances.

Estimating price fluctuations

In practice an employer will ask for a firm price or a price subject to fluctuations using a price adjustment formula.

Fluctuations: formula method

When a formula method is used for adjustment of costs the amount the contractor recovers from the employer is calculated by reference to predetermined rules. It can be considered as an adjustment in cost which has been deemed to have been experienced by the contractor. Most contractors who operate the formula fluctuations agree that the amounts recovered yield a reasonable recovery of the increased costs that occur.

It is important to note that a price adjustment formula option is not based on the actual changes in prices, but it uses the cost indices calculated from data collected from British manufacturers, and published monthly by central government. The indices therefore do not reflect the cost changes relating to goods which are manufactured abroad, and transport costs and currency changes are also at risk.

Items for which fluctuations are not considered are:

- Unfixed materials on site (price fluctuations will be included when the materials become part of the permanent works).

- Plant which is valued on a daywork basis (the plant schedule may become dated during a contract with a long duration).

- Claims (because they are calculated at their true value).

- Non-adjustable element under public forms of contract.

Any over- or under-recovery should be noted in the estimator's report for consideration at the final review meeting.

Firm price allowance

When a firm price tender is to be given, an assessment must be made of the likely variations in cost during the proposed contract period, including the period for 'acceptance of tender' stated in the contract documents. The tender programme is an important tool in assessing the likely impact of price rises on each element of the project. There are some parts of an estimate which do not need consideration such as provisional and PC sums, and firm price quotations which fully comply with the conditions of contract.

With the widespread use of analytical estimating (and the summary forms as shown in Figures 11.1 to 11.4) increased costs can be assessed for each resource and sub-contract package.

Labour

Each company must keep records of wage rates and plus rates due under the Working Rule Agreement as well as bonus earnings. Note must be taken of industrial negotiations and statutory measures which may affect levels of wages. If wage agreements are not known at the tender date, then it is recommended that discussions with the trade representative body should take place requesting guidance as to the likely outcome of such negotiations.

Materials and plant

Records must also be kept of the changing cost of materials and plant hire costs. From these records and detailed discussions with suppliers and manufacturers, the likely trends in prices can be more accurately assessed, rather than making an arbitrary guess at price increases.

There are three ways in which increased costs are calculated:

1. Each resource listed on the materials summary is considered in turn, depending on its expected purchase date.

2. An overall assessment can be made by picking a point on the programme when it is believed materials will be subject to an increase.

3. The contractor may decide to place early orders for high-value materials, or those subject to known increases. There could be an additional cost, however, in financing the purchases, particularly where the client will refuse to include unfixed materials in interim valuations.

Although it is a very difficult assessment to make, regard must be paid to economic and political situations at home and abroad and the state of both home and world markets.

Sub-contractors

Enquiries to sub-contractors must give clear instruction and information on the contract provisions for fluctuations together with an outline programme showing when the trade package will be carried out. The quotations from the contractor's own sub-contractors must be examined very carefully to ensure that each offer is for a firm price contract and for the period required by the contractor's invitation details and that there are no risks of claims for increased costs. The sub-contract summary form shown in Figure 11.4 has a column for assessing any shortfall.

Project overheads

Project overheads may suffer inflation, the most significant item being supervision. Staff costs are usually reviewed annually and the increases are to some extent predictable. Fluctuations in project overheads are dealt with on Figure 11.5, the 'Firm price adjustment' form.

11.4 FINAL REVIEW ADJUSTMENTS

When the resource summaries have been completed, it would be unwise to change them during the final review stage, for three reasons:

- Those reviewing the tender could become confused and waste time marking up their copies.

- Mistakes are easily made when summaries are changed during a meeting.

- The computer reports would no longer match the summary forms.

Figure 11.6 illustrates a typical adjustment form which enables the changes to be made under the usual headings of labour, plant, materials, sub-contracts and overheads. The last category, 'overheads', provides for certain aspects of project overheads which do not fall within the normal analysis.

The adjustments form also gives a full listing of the many decisions which affect the estimate. If the tender is successful, this form will be used as a checklist to adjust the rates and prices in the estimator's bill so that accurate data can be given to the site team as a budget for cost control.

11.5 VALUE-RELATED ITEMS

There are a number of costs, generally project overheads, which will be calculated as a proportion of the total value of work. Some bills of quantity provide items for these costs on the final summary page. They are:

- Contractor's insurances.

- Performance bond.

- Water for the works.

- Professional fees for design.

Each contractor has its own method for dealing with these items, either as project overheads or in the final summary. If the project overheads forms are used, an adjustment may be needed when the final tender price is known.

11.6 LATE QUOTATIONS

Any late quotations that have been received for the supply of materials, hire of plant or work of sub-contractors, must be entered on the appropriate summary sheets and comparison made with previously received quotations. The summary forms given in Figures 11.1 to 11.4 include columns for late quotations for the estimator to use before the final review meeting. Following the meeting, the estimator must use the final review adjustments form after consultation with the review manager. Great care is needed in comparing late quotations:

- There may be qualifications which other suppliers or sub-contractors may not have made.

- The items included in the quotation must be the same as those priced by others.

- The net amount of a late quotation must be compared with the previous lowest tender, including any adjustments made before or during the review meeting.

- It may be difficult to contact suppliers or sub-contractors to discuss queries about their quotations.

11.7 CHECKING PROCEDURES

In previous chapters, simple procedures are given to check that all the tender documents have been received, quotations are analysed, bills of quantity have been entered correctly in the computer and items have been correctly priced in accordance with the client's requirements. These are all parts of an overall quality system which an estimating department must adopt.

A quality assessment review can be undertaken at various stages but must be completed before a tender is submitted. Figure 11.8 is an example of a quality system checklist for the whole of the tender stage. It must be recognised that merely filling in a form does not constitute a check; there should be systems which enable the estimator to meet specific requirements. As a general rule, calculations should either be 'self-checking' by well-designed forms or someone else must repeat the calculation in order to eliminate mistakes. With computer-aided estimating systems, the number of calculation errors tends be less but other problems arise through miskeying data and partly pricing items. Completely unpriced items can be listed in a computer report.

Since most tenders are judged on price, it is important to check that the tender figure has been calculated correctly and is sufficient to meet the specific requirements of the client. A badly prepared tender could have the following effects:

- If it is under-priced there may be difficulty in achieving the specification for the price and there will be a potential loss of profit.

- If it is over-priced, the tender will not win the contract in competition.

- If it is badly priced, unreliable cost data would be provided for construction, and it would be difficult to achieve the desired financial outcome.

The total unit rates must be reasonable and correct for the unit of measurement. All calculations should be subjected to an arithmetical check. Particular attention must be given to the major items in each trade and to those items in which the materials specified are unusually costly. Any inaccuracy in the rate build-up for these items will be magnified and have very serious consequences.

It is equally important to check the sub-contractor-related items to see that items of work, plant and material not included in their quotation have been correctly assessed and incorporated in the rest of the item and that the unit rates reflect these factors.

It is important to ensure that when the time for a project is prescribed by the client, it is realistic in relation to the estimated labour cost and number of men required to be physically working on site.

The construction programme must be reviewed critically to ensure that no major problem has been overlooked and that the method statement represents the best method of completing the project. The

*Figure 11.8 **Quality system checklist***

KEYSTONE CONSTRUCTION LTD	QUALITY SYSTEM CHECKLIST		Tender number 97/201
Ref.	Description	Comments	Complete
1	Tender information form		✓
2	Check documents received		✓
3	Date stamp documents received		✓
4	Tender register	not used	
5	Acknowledge tender documents		✓
6	Requests for information	not required	
7	Estimating programme		✓
8	Tender timetable		✓
9	Distribute documents		✓
10	Site visit	23 July	✓
11	Computer input		✓
12	Enquiry documents		✓
13	Sub-contractor vetting		✓
14	Sub-contract comparisons		✓
15	PC and provisional sums		✓
16	Tender programme		✓
17	Unpriced items check		✓
18	Part-priced items check		✓
19	Project overheads		✓
20	Bill summaries		✓
21	Resource summaries		✓
22	Firm price allowances		✓
23	Cash flow	not required	
24	Confirm tender amendments	see tender letter	✓
25	Final tender review	4 August	✓
26	Submission documents	prog, priced bills	✓
27	Priced bills of quantity		✓
28	Tender letter		✓
29	Copy tender submission documents		✓
Notes			

estimator may be able to compare the estimate and final costs of previous projects of a similar nature in order to form an opinion on the likely profitability of the project.

11.8 FINAL SUMMARY

The final summary form (see Figure 11.7) is completed in stages by the estimator, before the final review meeting, as far as the first sub-total 'estimate frozen' line, and during the final review meeting in order to arrive at a tender price. It is important to identify the stage at which the estimate is 'frozen' because all later adjustments are made to these figures before the final tender figure is agreed. There is seldom sufficient time for the estimator to return to the computer system in order to incorporate the changes in the computer.

Provisional sums and daywork are added after the mark-up because they include an amount for profit.

Since this form is used by management to 'mark-up' a tender it is confidential and subject to a restricted circulation.

11.9 CASH FLOW

Cash is a major resource in construction and its use must be anticipated and managed. Production on construction projects must be planned and the effect upon resources and the return on assets employed determined. Clearly, the return on capital will be influenced by the speed at which it is earned. A simple cash flow diagram must be regarded as a useful aid to decision making and the estimator should ensure that such a diagram is produced for consideration at the final review stage.

When a contract has been secured, a fully-detailed forecast of cash flow, budget and cost control will be essential. However, post-contract action is outside the scope of this Code.

The estimator's build-ups will have established the costs of the work and project overheads under the headings of, labour, plant, materials, domestic sub-contractors, PC sums and provisional sums and daywork. In order to produce a cash flow diagram this financial appraisal must be linked to the tender programme. Cash resources must be established for each item on the programme. Once this link between estimate and programme has been established, the cash flow position can be calculated. The following headings must be taken into account when assessing the financial commitment to the project and the cash flow forecast:

- The contractor's own trading conditions

  — The anticipated cost commitment for own labour, plant, materials, sub-contractors and overheads.

- Contract conditions

  — Time period before issue of the first certificate.

  — Interval between certificates.

  — Anticipated time of practical completion.

  — Anticipated time for ending of defects liability period.

  — Delays between issuing of certificates and receipt of cash.

  — Retention during construction.

  — Maximum retention (for civil engineering contracts).

  — Retention after practical completion.

  — Period for final measurement.

- Banking conditions

  — Earning rate of interest.

  — Paying rate of interest.

Other factors which may also need to be considered include:

- Timing of special payments, for example, special insurances.

- Bonds, including date of release.

- The allocation of firm price allowances.

- The recovery of fluctuating costs.

- An efficient use of experienced staff – good staff can be allocated to projects bringing a poor return, whereas they may have been employed in situations which would yield better margins.

- Any special weighting that may be incurred on the contract.

- Provisions for liquidated damages.

- Any purchases of plant or equipment.

- The present financial commitment of the company and effect of the current tender on that commitment.

- Turnover and commitment of resources.

- Statutory requirements; for example, VAT.

- Level of project overheads and means of recovery of cost.

- Amount of PC and provisional sums and their likelihood of being spent.

- Company's prior experience of the client and his consultants.

- The addition to be made to the net cost estimate for head office overheads and profit and the manner in which this addition is to be made.

A cash flow calculation will indicate the financial support needed for the contract and the financial contribution to be made to the company.

When a client requests a cash flow forecast, at tender stage, the contractor is asked to submit a schedule of gross valuations at the date of each interim certificate, based on the tender programme. Such a document would have no contractual standing but assists the client to prepare a funding schedule.

The calculations needed to produce a cash flow forecast can be complicated and would take time which an estimator can rarely afford. In particular, the contract value graph is based on income from rates contained in the priced bills of quantity which are produced at the end of the tender period. Computers can produce the data and charts, using either a computer-aided estimating system or (partly written) spreadsheets.

12. FINAL REVIEW AND TENDER SUBMISSION

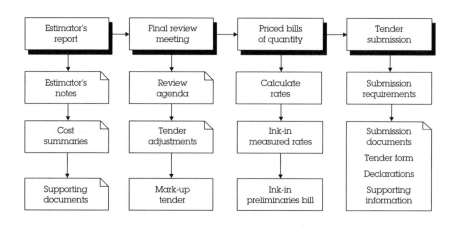

12.1 ESTIMATOR'S REPORT

The estimator's report brings together all pertinent facts which have influenced the preparation of an estimate for final review by management. The object of this report is to highlight to management the various matters which have been identified as cost significant, where alterations have been made to normal production standards and any special or unusual contract conditions or risks. These matters are described more fully below. An example of a form which can be used for the estimator's report is shown in Figure 12.1. It should be noted that this is only one representation of data to be considered at this stage. The way in which information is presented will vary according to company policy and preference, and the guidelines suggested should not be regarded as mandatory.

Many of the resource schedules, project overheads and other summaries are produced by computer-aided estimating software, often tailored to a company's procedures. Whichever system is used, the estimator's report to management should include:

- A description of the site and its location.

- A brief description of the project.

- A description of the method of construction.

- A note of any actual risks which are inherent in the project.

- Any information regarding the client, architect, quantity surveyor, consultants or other member of the professional team that should be brought to the attention of management.

Figure 12.1 Estimator's report

KEYSTONE CONSTRUCTION LTD	**ESTIMATOR'S REPORT**	Project 97/201

Supporting documents

Agenda	X	Method statements	☐	Schedule of PC sums	X
Tender information form	X	Tender programme	X	Schedule of prov sums & daywork	X
Site visit report	X	Cash flow analysis	☐	Project overheads	X
Materials comparison form	X	Estimator's notes	X	Resource summaries	X
Plant quotations register	X	Drawings	X	Final summary	X
Domestic sub-contractor's register	X	Specifications	X		

Conditions of contract

1. Design portion agreement required for precast concrete floors

2. Phased handover may be required but no sectional completion clauses available

3. Not clear when the performance bond will end

Risk

1. Not allowed trench sheeting to manholes next to workshops

2. Difficult to interpret full scope of work for demolition of stores and basement

3. Concrete supplier used cement replacement in some mixes

Scope

1. Topsoil to be taken off site may be sold, saving approximately £12,300

2. Alternative steel doors from KTL Doors would save £2,300

3. Quantity of concrete in foundations may be more than billed

Competition

LL Upman (London)   Duke and Dunfiled

Redchurch Developments   MAL Construction

Coinbrook Construction

Notes

- The conditions of contract.

- Employer's special conditions, such as bond and insurances.

- Any unresolved technical or contractual problems.

- An assessment of the state of the design and the possible financial consequences thereof.

- A note of any major assumptions made in the preparation of the cost estimate.

- An assessment of the profitability of the project.

- Any pertinent information concerning market and industrial conditions.

- Any need for qualification of the tender or for an explanatory letter.

- The terms of quotations from own sub-contractors which have been included in the estimate.

- The time for which the tender is to remain open for acceptance.

- Details of other tenderers where known.

- Future prospects arising from the scheme such as 'follow-on' work.

Estimate schedules and summaries

The estimator's report will include documents which have been produced during the preparation of the estimate and summaries relating to the cost estimate, including:

- Schedule of prime cost sums and attendances.

- Schedule of provisional sums and daywork.

- Materials comparison form.

- Plant quotations register.

- Domestic sub-contractors' register.

- Resource summaries for labour, plant, materials and domestic sub-contractors.

- Project overheads.

- Final summary.

Supporting documents

The cost summaries will in turn be supplemented by supporting documents which will provide the technical background, including:

- Method statements.

- Tender programme.

- Cash flow calculations.

- Tender drawings.

- Specifications and ground investigation report.

- Quotations files for: domestic sub-contractors, materials and plant suppliers.

- Tender information form.

- Site visit report.

This information will then be considered by management at the final review stage.

12.2 FINAL REVIEW

The final review of an estimate and its conversion to a tender is the responsibility of management and is a separate commercial function based upon the cost estimate and its supporting reports and documents.

The accountability of the estimator should be limited to the proper preparation of the predictable cost of a project. It must not be considered that the estimator's responsibility is to secure work for the company; that is the responsibility of management.

The final review of a construction estimate and conversion into a tender is similar in thought and method to any other manufacturing industry; the need to secure work in financial and production terms is paramount to the success of any business.

Final review meeting

Although the final decisions are made by management, those concerned with estimating, planning, management and buying must be encouraged to communicate the knowledge they have acquired throughout the estimating stage to the review panel. Their contribution may be by attending the meeting or reporting through the estimator.

In larger companies, the final review stage often consists of two meetings: the first to review the estimate through a detailed examination of rates and quotations; the second would be for a director to receive a short briefing and consider commercial matters before settling on a mark-up. The need for a formal approach finalising tenders should be regarded as fundamental to competent tendering.

An agenda for review meetings is indispensable until such time as orderly, logical and methodical thinking becomes second nature. The standard review agenda given in Figure 12.2 can be used for most projects. For larger schemes, the agenda must be circulated to those required to attend, with times shown for each agenda item.

An evaluation of alternatives, scope to improve profitability and risks which may be encountered will be considered at each stage of the estimating process and converted into costs at the final review meeting. Clearly, in a strong competitive market, contractors need to find every opportunity to use products and processes which will enable the contract to support sufficient overheads and profits to maintain the company's objectives and satisfy the client by completing on time with the required specification.

The mark-up shown on the final summary form (see Figure 11.7, page 161) comprises scope, risk, head office overheads and profit. Mark-up will be applied to all costs above this point, and it is assumed that provisional sums and daywork will produce their own contribution to overheads and profit.

Figure 12.2 Standard review agenda

KEYSTONE CONSTRUCTION LTD	**STANDARD REVIEW AGENDA**	Project	
The site	1 Location and ground conditions		
	2 Project description and alternatives		
	3 Hazards and security		
	4 Site visit report and photographs		
The parties	5 Client		
	6 Consultants		
The contract	7 Form of contract		
	8 Amendments		
	9 Bonds		
	10 Insurances		
	11 Damages		
	12 Payments and retention		
The programme	13 Method statements		
	14 Specific requirements		
	15 Tender programme		
The estimate	16 Labour		
	17 Plant		
	18 Materials		
	19 Sub-contractors		
	20 PC and provisional sums		
Project overheads	21		
Final summary	22 Net estimate		
	23 Firm price adjustment		
	24 Review adjustments		
	25 Value-related items		
Mark-up	26 Cash flow		
	27 Scope		
	28 Risk		
	29 Qualifications		
	30 Head office overheads		
	31 Competition		
	32 Profit		
	33 Discounts (for main contractor)		
	34 VAT (where applicable)		

At the end of the final review meeting, there may be other decisions to be made before proceeding to the submission stage, including:

- Which documents will be submitted with the tender? Some contractors may wish to prepare additional information such as a programme, company profile and coloured brochures, whereas others reserve their efforts, and ideas, until they know that their tenders are under serious consideration.

- If an alternative method (or design) is to be offered, then another price may need to be settled by the review panel.

- Qualifications, which vary the requirements of the tender documents, are generally not permitted by clients. On the other hand, there may be circumstances in which they cannot be avoided, and management must decide the form that a qualified tender will take.

- In order to provide a positive cash flow for the duration of the project it is important to decide how rates are to be apportioned in a bill of quantity to be submitted at tender stage? Since any artificial alteration of prices will bring additional risks, it is for management to agree the strategy.

Records

It is recommended that a record is kept of all tenders submitted. Particular care should be taken of the final summary form, and all notes and details of the decisions taken at the review meeting should be fully recorded for future reference.

12.3 PREPARATION OF PRICED BILLS OF QUANTITY

At the final review stage, management may alter some of the basic conceptions concerning the project. For example, a change to the basic labour rate will affect all bill items containing labour and, therefore, affect all net rates that have been compiled by the estimator. With computer-aided estimating systems such changes can be easily accommodated at the end of the tender period but often there is insufficient time to rewrite the net rates when manual estimating systems are used.

If priced bills of quantity are to be submitted at the time of tender (or if requested later), management must decide how the difference between the total of the net rates in the draft bills and the agreed tender figure is to be shown. The method chosen will depend on individual preference.

Examples of possible methods are:

- All amendments being made to the respective elements of the tender. (This may mean substantial repricing of the bills of quantity, and a balance inserted in the preliminaries bill.)

- Unit rates increased by an agreed percentage so that the whole difference is included in the measured items.

- The unit rates remain net and the whole difference included as an adjustment on the final summary page – providing this is considered to be acceptable to the client.

- The unit rates remain net and the difference inserted in the preliminaries bill.

- Any combination of these methods.

The way in which the submission bills of quantity are priced will have a significant effect on income generated through interim payments and the valuation of variations. Clearly a contractor would not be able to insert highly inflated prices against early items because a client will not accept a disproportionate distribution of rates. The contractor also puts his position at risk if under-measured items are under-priced, and over-measured items are over-priced; in both cases where the work is remeasured, the contractor will incur a loss.

12.4 TENDER SUBMISSION

The estimator must ensure that the procedure set out in the tender documents for the submission of the tender is followed meticulously. The form of tender, and any other declarations, must be completed as required and signed by an authorised person from the contractor's organisation.

A letter should be attached to the forms to list the documents, and confirm amendments received during the tender period. Any other remarks about the technical content or price may be viewed as a qualification to a tender and so should be avoided. The codes of procedure for selective tendering, whether for main contractors or trade contractors, state that a tenderer who submits a qualified tender should be given reasonable opportunity to withdraw his qualifications without amendment of his tender price. Failure to do so will lead to rejection of the whole tender if it is considered that such qualifications give the tenderer an unfair advantage.

Alternative proposals involving design, specification or timing may be invited. In this case those contractors who wish to offer alternatives must submit a separate tender indicating the effect on price and programme, together with an appraisal of any changes needed to other elements of the scheme.

The NJCC *Code of Procedure for Single Stage Selective Tendering* (CPSSST) recommends that tenders should be opened as soon as possible after receipt. Submission of priced bills of quantity may be required with the tender, but in any event the CPSSST recommends that the lowest tenderer should be invited to submit priced bills of quantity within four working days of the opening of the tenders.

13. ACTION AFTER SUBMITTING A TENDER

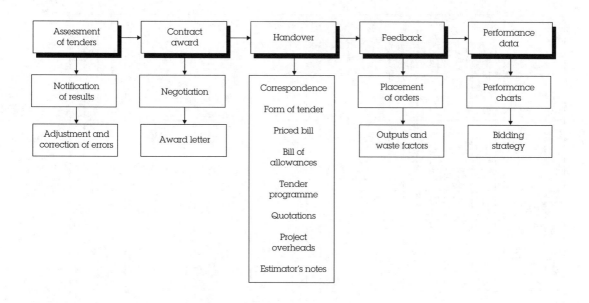

13.1 ASSESSING TENDERS AND NOTIFYING RESULTS

The *Code of Procedure for Single Stage Selective Tendering* recommends that tenders should be opened promptly, and all but the three lowest tenderers should be informed immediately that their tenders have been unsuccessful. The second and third lowest tenderers should be informed that their tenders were not the most favourable and that they might be approached again if it is decided to give further consideration to their offers. They must be notified at once when a decision to accept a tender has been taken.

Once the contract has been let, every tenderer must be promptly supplied with a list of the tender prices. The estimator (or chief estimator) should record the results in the tender register (see Figure 4.1, page 31) and, if required, in a tender performance report illustrated in Figure 13.1.

13.2 ADJUSTMENT OF ERRORS

The estimator should be prepared to respond to any request for further information or a notification that the submitted tender contains errors in computation. For errors, two alternative courses of action are possible:

TENDER PERFORMANCE REPORT

KEYSTONE CONSTRUCTION LTD

From 12 September

To

No.	Client	Title	Tender date	Our tender	Lowest tender	% below next	% over lowest	% over mean	Date of award	Comments
10051	Hamilton Systems	Factory extension	12 Sep	1,012,843	989,523		2.4	-2.3		Marsh Construction
10052	Direct Water	Water treatment works	21 Sep	525,810		-2.8			10 Nov	
10053	Thomas Insulation	Car park	10 Oct	185,231	143,100		29.4	not known		Hanplan Bros
10054	Home Office	Hanfield prison	12 Oct	1,920,500	1,816,250		5.7			Marsh Construction
10055	RSPCA	New animal shelter	27 Oct	525,953						
10056	Campton DC	Harlow primary school	05 Nov	844,252						negotiated
10057	Saturn Developments	Victoria House	10 Nov	2,565,100						

Figure 13.1 Tender performance report

- The tenderer will be given details of such errors and afforded the opportunity of confirming or withdrawing the offer. The estimator will need to refer to management when the extent of the computation errors have been determined for a decision as to whether to confirm the original tender figure or withdraw the tender. Under English Law it is possible to withdraw the tender at any time before its acceptance.

- The second option is that the tenderer is given an opportunity of confirming his offer or of amending it to correct genuine errors. If the contractor elects to amend his offer and the revised tender is no longer the lowest, the second lowest tender will be examined in more detail.

The estimator must consult with management to establish whether to amend the tender figure or to confirm the original offer, once the extent of the computation error has been determined.

In both situations the estimator or signatory to the original tender must be prepared to endorse the appropriate tender documents to note the acceptance or change to the tender. Such amendments must also be endorsed by the employer in the event that a contract is subsequently awarded.

There are many ways to correct errors in a bill of quantity. The most common is to recalculate the bill and make an adjustment on the final summary page, either by changing the total for preliminaries or introducing an adjustment item. Since the latter amount must be applied to all valuations, it is preferable to ask the contractor to change one or more items in the preliminaries to bring the total to the tender figure.

13.3 NEGOTIATION AND AWARD

When the lowest tender received exceeds the client's budget, changes should be negotiated with the lowest tenderer. This process is either through recommendations from the contractor for cost savings, or design changes which reduce the scope or specification of the works. If significant changes are proposed to a scheme, two or (at most) three tenderers may be asked to re-tender in competition, thereby retaining the lowest market prices for the amended project.

If there has been any delay in acceptance, enquiries should be made to ensure that there have been no changes which might necessitate reconsideration of the tender.

Once an acceptable offer has been made, the award of a contract should be in clear terms stating that the offer has been accepted. There should be no reason to use a letter of intent which can lead to confusion about how payments will be made for abortive work. Whichever method is used, the estimator must obtain approval from management that the wording can be accepted in order to start planning and mobilisation of resources. A standard contract award form, such as that shown in Figure 13.2, should be used to provide an overview for other departments and a facility for the responsible director to indicate his acceptance of the terms and provide the authority for the construction team to incur expenses.

13.4 ACTION WITH A SUCCESSFUL TENDER

When a tender has been successful it is necessary to take certain action before the actual contract documents can be signed. This action will be carried out by various people within the company's organisation but it is recommended that it should be co-ordinated by the estimator who was responsible for the original submission. The documents must be checked thoroughly to ensure that they reflect the exact content of the documents used to prepare the tender.

Contract documents

Contract documents should be checked to establish that:

- The drawings are those which were circulated with the tender documents; if they have been revised to produce a construction issue, they cannot be used as contract documents.

- The dates, penalties and particulars given in the appendix to the conditions of contract are those stated in the tender document.

- The submission bills have been copied correctly.

A standard form may be used to confirm that these checks have taken place. In larger organisations, this form will accompany the documents until they are signed by the company's directors and returned to the client's representative.

Figure 13.2 Contract award

KEYSTONE CONSTRUCTION LTD	**CONTRACT AWARD**	Tender number Contract number

Project particulars

Project title		Client	
		Project manager	
Project address		Architect	
		Quantity surveyor	
Project description		Structural engineer	
		Services engineer	
Award letter	Date	Planning supervisor	
Acceptance			
Letter of intent		Principal contractor	
Other			

Contract

Start		Contract value	
Completion date		Bond Value	
Duration		Release date	

Early action

External	Date		Internal	Date	
Acknowledgement letter			Mobilisation programme		
Pre-start meeting			Health and safety plan		
			Handover meeting		
			Start-up meeting		

Distribution

☐ Commercial	☐ Operations director	
☐ Purchasing	☐ Area manager	
☐ Insurance	☐ Contracts manager	
☐ Accounts	☐ Business development Marketing	
☐ Safety	☐	PR
☐ QA	☐	

Signed   Date

(Director)

Handover: distribution of information

The following information, as appropriate, must be made available immediately to those who will be responsible for construction and purchasing:

- Correspondence.

- Form of tender.

- Priced submission bill.

- Bill of internal allowances.

- Build-up of rates fully adjusted for review changes.

- Tender programme and method statements.

- Quotations.

- Analysis of quotations.

- Project overheads.

- Estimator's notes.

- Site visit report.

- Further information received after tender submission (if any).

- Tender documents.

It is recommended that an internal pre-contract 'handover' meeting be held with all persons concerned in attendance so that details are fully discussed and the distribution of information can take place at the meeting. The handover meeting is an opportunity for detailed reporting and a discussion on decisions made at the time of estimating concerning methods of construction, site layout, organisation, suppliers and sub-contractors. These decisions should never be made binding on those to be responsible for construction. It is important that the reasons underlying such decisions and choices should be explained fully. However, decisions made at the time of tender cannot be ignored unless it can be demonstrated that a better method of operation is possible.

For a detailed appraisal of handover procedures see the *Code of Estimating Practice*, Supplement No.4: *Post-tender Use of Estimating Information* published by the Chartered Institute of Building.

13.5 FEEDBACK INFORMATION

It is also recommended that a fully operational cost control system be introduced on all successful contracts. Although it is outside the scope of this Code it should be noted that the collection and collation of data is a specialised function involving the use of statistical techniques. It is important that during construction information should be obtained on the labour and plant cost of major items or stages of work, on the quantities of material used and on the cost of attendance for sub-contractors. This information is essential for control purposes and guidance in future estimating. These records should be accurate and give all details including those of the work, the position on the site of the operation, gang sizes, levels of incentives and bonuses being paid, plant, access and weather conditions. This information is needed so that it may be interpreted correctly for future use.

Where a substantial proportion of work is assigned to sub-contractors, a table should be drawn up showing the estimator's allowances and actual prices agreed with them at the time of placing their contracts. In particular, the estimator needs to know the accuracy of risk, buying gains and firm price provisions in the tender.

A brief report should also indicate the performance of all sub-contractors and a recommendation on whether they should be employed for future projects.

Visit to the site

Those who were responsible for preparing the estimate should make periodic visits to the site during the progress of construction in order to:

- Assess objectively the soundness of decisions made during the estimating process or subsequent site decisions where changes were considered necessary to improve performance and/or outputs.

- Maintain a constant awareness of the conditions which prevail on sites.

Final reconciliation

At the completion of a project, the estimate and tender should be reconciled with the final cost and final account and the reasons for the results assessed in detail for guidance in future estimating.

13.6 ACTION WITH AN UNSUCCESSFUL TENDER

When the result of a tender is known the tender performance form (see Figure 13.1) should be completed, where possible with a comparison made with the lowest (or the accepted) tender. Tender results can be reported at regular management meetings so that all those who contributed to the estimate can assess their performance. Suppliers and sub-contractors who submitted quotations should be notified of the results as promptly as possible.

Tender documents should be carefully filed for possible future reference – except drawings and bills of quantity which are of little use once the contract has been awarded to another contractor. Quotations received from suppliers and sub-contractors can provide a useful guide to prices for other tenders *but great care is needed* to ensure that the specification and the nature of the work are the same.

13.7 TENDER PERFORMANCE

Estimators must monitor their performance with an analysis of results recorded over a period of consistent estimating. If data are available from a large number of tenders, it is possible to evaluate tender performance in relation to: types of work (new build or refurbishment), clients (public, private), procurement routes (traditional, design and construct, trade packages) and value of projects. With this information, decisions can be made about where to concentrate estimating resources for further tendering.

13.8 BIDDING STRATEGY

A contractor, or specialist trade contractor, will formulate aims and a framework to win contracts at prices which will produce profitable work. An evaluation of tendering performance in itself is not enough to meet this aim, because although a contract may be won with a reasonable anticipated profit, it may be found that the particular type of work seldom achieves or improves margins when the final results are known.

A bidding strategy must recognise the need to decline invitations to tender for work or procurement arrangements which fail to meet the organisation's objectives.

When contractors set tender margins, they consider all the points discussed in the Final Tender Review meeting (see Figure 12.2, 'Review agenda', on page 177). In addition, they may be influenced by tenders submitted by their competitors and budgets set by clients. It is clear that neither of these criteria will help achieve optimum margins in a highly competitive market. It is difficult to avoid low margins where competition is strong but, if prices fall below economic levels, it is time to look for other markets where the company can employ its strengths more effectively.

REFERENCES FOR FURTHER READING

Aqua Group (1990) *Tenders and Contracts for Building*, 2nd edition, Blackwell

Aqua Group (1992) *Pre-contract Practice for the Building Team*, 8th edition, BSP

Ashworth, A. (1994) *Cost Studies of Buildings*, 2nd edition, Longman Scientific and Technical

Ashworth, A. and Skitmore, R.M. (1982) *Accuracy in Estimating*, Occasional Paper No. 27, The Chartered Institute of Building

BEC/RICS (1988) *Standard Method of Measurement of Building Works*, 7th edition

Brook, M. (1993) *Estimating and Tendering for Construction Work*, Butterworth Heinemann

Buchan, R.L., Fleming, F.W. and Kelly, J.R. (1991) *Estimating for Builders and Quantity Surveyors*, Butterworth Heinemann

Civil Engineering Standard Method of Measurement, 3rd edition (1991), Thomas Telford

Co-ordinating Committee for Project Information (1987) *Co-ordinated Project Information for Building Works: a guide with examples.* (This and other CPI documents may be obtained from BEC Publications, RIBA Publications or Surveyors' Bookshop.)

Cook, A.E. (1991) *Construction Tendering: theory and practice*, Batsford

Cross, D.M.G. (1990) *Builders' Estimating Data*, Heinemann Newnes

Department of the Environment PSA (1987) *Significant Items Estimating* (produced by the Directorate of Building and Quantity Surveying Services), HMSO

Farrow, J.J. (1984) *Tendering – An Applied Science*, 2nd edition, The Chartered Institute of Building

Franks, J. (1996) *Building Procurement Systems*, The Chartered Institute of Building

Kwakye, A.A. (1994) *Understanding Tendering and Estimating*, Gower

Latham, Sir Michael (1994) *Constructing the Team* (joint review of procurement and contractual arrangements in the United Kingdom construction industry), HMSO

McCaffer, R. and Baldwin, A. (1991) *Estimating and Tendering for Civil Engineering Works*, 2nd edition, BSP

Ministry of Public Building and Works (1964) *The placing and management of contracts for building and civil engineering work* (Report of the committee, Chairman: Sir Harold Banwell), HMSO

Ministry of Works (1944) *The placing and management of building contracts* (Report of the committee, Chairman: Sir Ernest Simon), HMSO

Sher, W. (1997) *Computer-aided Estimating: a guide to good practice*, Addison Wesley Longman

Skitmore, R.M. (1989) *Contract Bidding in Construction: strategic management and modelling*, Longman Scientific & Technical

Smith, A.J. (1995) *Estimating, Tendering and Bidding for Construction*, Macmillan

Smith, R.C. (1986) *Estimating and Tendering for Building Work*, Longman

The Chartered Institute of Building (1978) *Information Required before Estimating: a code of procedure supplementing the Code of estimating practice*

The Chartered Institute of Building (1981) *The Practice of Estimating* (compiled and edited by P. Harlow)

The Chartered Institute of Building (1983) *Code of Estimating Practice*, 5th edition

The Chartered Institute of Building (1987) *Code of Estimating Practice*, Supplement No. 1: *Refurbishment and Modernisation*

The Chartered Institute of Building (1988) *Code of Estimating Practice*, Supplement No. 2: *Design and Build*

The Chartered Institute of Building (1989) *Code of Estimating Practice*, Supplement No. 3: *Management Contracting*

The Chartered Institute of Building (1993) *Code of Estimating Practice*, Supplement No. 4, *Post-tender Use of Estimating Information*

The Chartered Institute of Building (1996) *Code of Practice for Project Management for Construction and Development*, 2nd edition

Whitehead, G. (1990) *Guide to Estimating for Heating and Ventilating Contractors*, HVCA

CIOB Technical Information Papers, 1982–1991

Paper No. 7   Potter, D. and Scoins, D. (1982) *Computer-aided estimating*

Paper No. 9   Holes, L.G. and Thomas, R. (1982) *A general-purpose data processing system for estimating*

Paper No. 11   Harrison, R.S. (1982) *Practicalities of computer-aided estimating*

Paper No. 15   Skoyles, E.R. (1982) *Waste and the estimator*

Paper No. 37   Harrison, R.S. (1984) *Pricing drainage and external works*

Paper No. 39   Braid, S.R. (1984) *Importance of estimating feedback*

Paper No. 59   Uprichard, D.C. (1986) *Computerised standard networks in tender planning*

Paper No. 64   Ashworth, A. (1986) *Cost models – their history, development and appraisal*

Paper No. 65   Ashworth, A. and Elliott, D.A. (1986) *Price books and schedules of rates*

Paper No. 75   Harrison, R.S. (1987) *Managing the estimating function*

Paper No. 77   Ashworth, A. (1987) *General and other attendance provided for sub-contractors*

Paper No. 81   Ashworth, A. (1987) *The computer and the estimator*

Paper No. 97   Brook, M.D. (1988) *The use of spreadsheets in estimating*

Paper No.113   Senior, G. (1990) *Risk and uncertainty in lump sum contracts*

Paper No.114   Emsley, M.W. and Harris, F.C. (1990) *Methods and rates for structural steel erection*

Paper No.120   Cook, A.E. (1990) *The cost of preparing tenders for fixed price contracts*

Paper No.127   Harris, F. and McCaffer, R. (1991) *The management of contractor's plant*

Paper No.128   Price, A.D.F. (1991) *Measurement of construction productivity: concrete gangs*

Paper No.131   Brook, M.D. (1991) *Safety considerations on tendering – management's responsibility*

CIOB Construction Papers, since 1992

Paper No. 2   Massey, W.B. (1992) *Sub-contractors during the tender period – an estimator's view*

Paper No. 11   Hardy, T.J. (1992) *Germany – a challenge for the estimator*

Paper No. 16   Milne, M. (1993) *Contracts under seal and performance bonds*

Paper No. 17   Young, B.A. (1993) *A professional approach to tender presentations in the construction industry*

Paper No. 19   Harrison, R.S. (1993) *The transfer of information between estimating and other functions in a contractor's organisation – or the case for going round in circles*

Paper No. 23   Emsley, M.W. and Harris, F.C. (1993) *Methods and rates for precast concrete erection*

Paper No. 32   Moores, R. (1994) *Negotiation in construction*

Paper No. 33   Harrison, R.S. (1994) *Operational estimating*

Paper No. 49   Borrie, D. (1995) *Procurement in France*

Paper No. 54   Pokora, J. and Hastings, C. (1995) *Building partnerships*

Paper No. 55   Sher, W. (1995) *Classification and coding of data for computer-aided estimating systems*

Paper No. 64   Rossiter, J. (1996) *Better value from benchmarking*

The National Joint Consultative Committee for Building (RIBA and BEC)

Code of Procedure for Single Stage Selective Tendering (1994)

Code of Procedure for Two Stage Selective Tendering (1994)

Code of Procedure for Selective Tendering for Design and Build (1995)

Code of Procedure for the Selection of a Management Contractor and Works Contractors (1991)

Code of Procedure for the Letting and Management of Domestic Sub-contract Works (1989)

Guidance note 1   *Joint venture tendering for contracts in the United Kingdom* (1994)

Guidance note 2   *Performance bonds* (1986)

Guidance note 3   *Fire officers recommendations* (1993)

Guidance note 4   *Pre-tender meetings* (1988)

Guidance note 5   *Reproduction of drawings for tender purposes* (1990)

Guidance note 6   *Collateral warranties* (1992)

Guidance note 7   *Alternative dispute resolution* (1993)

Guidance note 8   *Construction management: selection and appointment of construction manager and trade contractors* (1994)

Guidance note 9   *Charges for admission to approved and select lists and for tender documents* (1995)

INDEX

absenteeism, 89
abstracting, 39
access to the site, 28
accommodation, 138
activity resources, 103
adequacy of information, 115
adjacent buildings, 44
adjustment
  item, 183
  of errors, 181
adjustments to estimate, 100
agenda
  final review, 176–7, 189
  interview, 15,16
  mid tender-review, 77
acknowledgement of tender
  documents, 18
all-in rates for labour, 79–82
  summary, 90–91
allocation of costs, 99
  comparison form, 65
  pricing, 98
  purchase, 56
  quotations register, 65–6
  schedules, 77
  summary, 77, 156
  suppliers, 56
alteration work, 23
alternative
  estimating techniques, 106
  methods, 75
  offers, 63, 65, 67
  tenders, 70, 178, 180
amendments, 22
analysis of quotations, 102
analytical estimating, 92, 163
annual
  employment costs, 85
  holiday stamp, 89
  holidays, 84
approved
  contractors, 7
  lists, 6–7
approximate
  cost, 8
  estimating, 106
  quantities, 109, 111
arithmetical checks, 167
assessing tenders, 181
attendance, 67
attendances, 62, 116
  nominated sub-contractors,
    115, 117
attendant labour, 148
attraction money, 82, 152
award of contract, 183–4

banksmen, 99
basic
  outputs, 93
  wages, 85
bidding strategy, 188

bill of internal allowances, 186
bills of quantity, 21, 28, 38, 69
British Summer Time, 84
budgets, 106
buildability, 39
builder's work, 122
buying department, 48

cash flow forecast, 169–71, 178
checking contract documents,
  184
checking procedures, 104, 166
chief estimator, 10, 21, 26–7, 30,
  32, 181
CITB training contributions, 81
civil engineering, 69, 77, 100, 110,
  126
clerical assistants, 105
client's
  brief, 107
  budget, 6, 106, 183
Co-ordinated Project Information
  (CPI), 28
co-ordination, 21, 30, 184
Code of Procedure for Single
  Stage Selective Tendering,
  10, 180–81
commencement, 73
Common Arrangement of Work
  Sections (CAWS), 38
common plant, 149
company-owned plant, 57
comparison forms, 63
competition, 5, 189
competitive
  advantage, 70, 75, 133
  tenders, 6
competitiveness, 75
completing cost estimates, 153
composite items, 112
computer
  assistants, 46
  estimating systems, 153
  input, 45
    checklist, 46–7
computer-aided estimating, 46,
  69, 79, 80, 93, 100, 103–4,
  112, 154, 178
  cash flow, 171
  operational estimating, 111
  systems, 63, 65, 67, 167
computers, 45
conditions of contract, 26, 37, 41
constants, 93
Construction Industry Training
  Board (CITB), 88
construction programme, 75
contingency sum, 123
contract
  award, 184–5
  conditions, 144, 150
  documents, 35, 45, 184

duration, 133
contractor selection, 6
contractor's
  bill of quantity, 112
  bonus, 81, 85
  cash flow, 26
  general cost items, 132
  presentation, 16
  tax exemption certificate, 62
contracts specialist, 26
correction of errors, 183
cost
  plan, 106
  planning, 112
  savings, 183
  yardsticks, 107

daily travel allowance, 82
daywork, 124, 126
  calculating percentage
    additions, 127–8
  materials and goods, 129
  materials discounts, 129
  overheads and profit, 126, 129
  percentages, 126
  plant, 130
    charges, 129
  prime cost of labour, 127
  schedule, 125
  sub-contractor, 126
Death Benefit Scheme, 89
decimal constants, 93, 103
decision to tender, 18, 27
delivery
  programme, 54
  times, 114
demolition work, 44
design, 28
despatch of enquiries, 35
detailed drawings, 23, 45
discounts, 63, 65, 67, 102, 105, 154
  materials prices, 96
  nominated suppliers, 114
discrepancies, 38, 63, 65, 67
  sub-contractor's quotation, 102
distributing materials, 148
distribution of documents, 40
divergence, 39
domestic sub-contractors, 59, 67
  register, 67–8
drawing lists, 32
drawings, 21–2, 38
drivers, 148

effective rate, 93
electricity supply, 150
electronic mail, 46
elemental cost plan, 107–8
Employer's Liability insurance, 81,
  89
employer's requirements, 132,
  136, 146